P9-DXT-523

THE CARBON CODE

THE CARBON CODE

HOW YOU CAN BECOME A
CLIMATE CHANGE HERO

BRETT FAVARO

JOHNS HOPKINS UNIVERSITY PRESS • BALTIMORE

© 2017 Johns Hopkins University Press
All rights reserved. Published 2017
Printed in the United States of America on acid-free paper
2 4 6 8 9 7 5 3 1

Johns Hopkins University Press
2715 North Charles Street
Baltimore, Maryland 21218-4363
www.press.jhu.edu

Library of Congress Cataloging-in-Publication Data

Names: Favaro, Brett, 1985–, author.
Title: The carbon code : how you can become a climate change hero / Brett Favaro.
Description: Baltimore: Johns Hopkins University Press, 2017. |
Includes bibliographical references and index.
Identifiers: LCCN 2016040198| ISBN 9781421422534 (hardback : alk. paper) |
ISBN 1421422530 (hardcover) | ISBN 9781421422541 (electronic)) |
ISBN 1421422549 (electronic)
Subjects: LCSH: Climate change mitigation. | Carbon dioxide mitigation. |
BISAC: SCIENCE / Environmental Science. | NATURE / Environmental
Conservation & Protection. | SCIENCE / Earth Sciences /
Meteorology & Climatology.
Classification: LCC TD171.75 .F38 2017 | DDC 363.738/746—dc23
LC record available at https://lccn.loc.gov/2016040198

A catalog record for this book is available from the British Library.

*Special discounts are available for bulk purchases of this book. For more information,
please contact Special Sales at 410-516-6936 or specialsales@press.jhu.edu.*

Johns Hopkins University Press uses environmentally friendly book
materials, including recycled text paper that is composed of at least
30 percent post-consumer waste, whenever possible.

CONTENTS

Acknowledgments   vii

Introduction   1

PART I. CLIMATE CRISIS   5

Chapter 1. The Cost of Carbon   7

Chapter 2. Solutions Start with You   32

Chapter 3. The Carbon Code of Conduct   47

PART II. LIVING BY THE CODE   63

Chapter 4. Electricity   65

Chapter 5. Transportation   103

Chapter 6. Adopting a Low-Carbon Diet   125

Chapter 7. Long-Range Travel   143

PART III. SHARING THE CARBON CODE   159

Chapter 8. Winning the Conversation   161

Chapter 9. Policies for a Pro-Climate Future   180

Chapter 10. Bringing It All Together   204

Notes   213

Index   215

ACKNOWLEDGMENTS

I thank my wife Corinna Favaro for her boundless support. I could never have completed this project without her.

I drew inspiration from my colleagues in the world of conservation science. Thanks to my colleagues, graduate students, and friends and family who have challenged me to be the best conservationist I can be.

Finally, I thank the team at Johns Hopkins University Press for giving me the opportunity to produce this piece of work and disseminate it to the world.

THE CARBON CODE

INTRODUCTION

There are few problems more intractable than climate change. It is a byproduct of our industrial success. The better our economies perform, the more energy they consume. And when we consume energy, that usually means we emit greenhouse gases, the gases that trap heat in the atmosphere and cause climate change.

There is no other problem like it. Climate change threatens to set back decades of development and impact every aspect of society. Every product we own, every chemical we use, and every kilometer we move is tied to fossil fuels. Every bite we take has a carbon cost, and every time we step on an airplane we add to the global greenhouse effect. It's death by a thousand cuts on a planetary scale.

But what does all this mean for you? You're an individual just trying to go about your business and lead a full, happy life. You want to be part of the solution but may fear that your existence makes you part of the problem. You recycle, you take the bus, and you feel you hold all the most progressive values related to the environment. But what can you do, as one person, that can tangibly alleviate the climate problem?

To put a finer point on it, as citizens of developed countries, how can we reconcile our desire to save the planet from the worst effects of climate change with our dependence on the systems that cause it? How can we demand that industry and governments reduce their pollution, when ultimately we are the ones buying the polluting products and contributing to the emissions that harm our shared biosphere?

These are the central questions of this book. My goal is to establish a set of rules that we can use to guide our decisions and make each and every one of us the protagonist of our own struggle against climate change. Whether you're a private individual just living your life or a prime minister or president looking for advice on how to run your country in a climate-friendly way, this book is for you.

If you consider yourself a conservationist or environmentalist, this book is definitely for you. We need to lead by setting the best possible example, and this book will give tangible strategies for how to do that.

If you have heard about climate change but only have a vague idea of what it all means, then this book will help give you a basic understanding of the issue. I hope that it also introduces you to a world of sustainable living that will help you become a solution to the climate problem. It will give you a better comprehension of some of the public debates that relate to the problem of climate change. You don't have to be a full-blown conservationist to take steps that can help you live a far more sustainable lifestyle.

If you think climate change is fake, that it's not caused by us, or that it won't be dangerous, then I hope you'll keep an open mind and read the book anyway. You'll probably get pretty angry reading certain sections, but I hope you will stick with it. The best part about fighting climate change is that many climate-friendly choices are also things that will save you money, improve your health, and make you less dependent on others for your well-being. You don't need to accept the science of climate change to understand that clean air, clean water, and renewable energy is worth a serious look.

There are three parts to this book.

In part I, I describe the carbon code of conduct, which is the core strategy that we will use to fight climate change. We start with some of the basic science behind the greenhouse effect, and I summarize some of the most serious problems that are currently being predicted by credible climate scientists (chapter 1). I make the case that you, as an individual, bear a responsibility to contribute to climate solutions and that your contribution can have a tangible impact on the problem (chapter 2). Then, I define the carbon code of conduct and explain the logic that underpins its design and structure (chapter 3).

But a carbon code is of no use without a way to follow it. Part II is all about this—in it, I cover the major activities of our daily lives that are responsible for the lion's share of an individual's typical carbon footprint. This section of the book is divided into electricity (chapter 4), short-distance transportation (chapter 5), the food we eat (chapter 6), and long-distance travel (chapter 7). In each, I identify the things we do that are particularly bad for climate change and show that in many cases there are perfectly good alternatives that can greatly reduce the amount of greenhouse gases released.

Whereas part I and part II focus on decisions that you can make as an individual, part III is designed to help you become a better climate advocate. It will cover how to win the conversation about climate change and how to constructively challenge anti-climate views and engage in the public debate in support of action (chapter 8). Then, in chapter 9, I go over nine public policies that the majority of experts agree we absolutely must implement to successfully stabilize our climate. Chapter 10 will conclude the book with some final messages about being a climate hero and avoiding the burnout that can occur when being involved in this struggle.

At the end of each chapter, you will find a short summary. This highlights the major points from each chapter that are the most important things to remember.

The mission of the carbon code of conduct is to build a system where we are able to productively exist in a society that uses more than its share of fossil fuels, while making decisions that reduce our footprint as much as possible. With enough people making these decisions, the ultimate goal of the carbon code can become a reality, facilitating a transition to a sustainable, carbon-neutral global economy.

To be clear, this book is not a comprehensive blueprint. It does not give the complete solution to the climate crisis, and it does not consider every caveat of every climate-friendly activity. Rather, it is meant to help you shape your behavior, so that the decisions you make are better for the biosphere than they were before. It is also meant to help you gain a basic understanding of the benefits of pro-climate policies—both in terms of how they make your life better directly but also in how they fight climate change. Finally, we explore some ideas on how to go about helping to get those policies enacted wherever you live.

The carbon code does not require perfection. You do not have to choose between being a perfect environmentalist or a terrible polluter. Rather, the goal must be for us to form a habit of making climate-friendly decisions in our day-to-day lives. Rather than being restrictive, the carbon code is empowering.

A quote by the thirteenth-century Persian poet Jalāl ad-Dīn Muhammad Rūmī sums up the philosophy underpinning the carbon code:

Yesterday I was clever, so I wanted to change the world. Today I am wise, so I am changing myself.

Now let's get to it.

PART I

CLIMATE CRISIS

Before we decide to save the planet, we have to define terms. What exactly do we mean by the term *climate change,* and why do we need a code of conduct to fight it? Why is our planet heating up, and why is that a bad thing? How serious is the climate problem, and is there anything we can do to stop it? Do we even matter, or is this problem bigger than we are?

In part I, we explore these questions. The first chapter outlines the basic science of climate change and demonstrates that this is a really big problem for life on this planet—human and otherwise. In the second chapter, we look at the role that each of us plays in harming the climate, how we bear a responsibility for healing it, and how each and every one of us can actually have a major impact on the climate problem. In the third chapter, I argue that we need to define and live by a carbon code of conduct to help us coordinate our action against climate change.

THE COST OF CARBON

I f you picked up this book, you probably believe that climate change is real. If you're like 99.9% of scientists, you also believe that it is caused by us, as a byproduct of the many industrial activities that power our lives. But a third fact, which is comparatively less understood, is that science is also telling us that climate change is basically one of the scariest things imaginable. Its effects are so serious and so far reaching that we have to do everything in our power to stop it. This book isn't about saving the planet. It's about saving ourselves.

Let's start with the basics. The atmosphere, or the blanket of gases that covers our entire world, is the planet's HVAC system. It helps us stay warm by trapping heat energy from the sun, preventing it from being lost out into space. The atmosphere also keeps us cool and provides a temperature buffer so that we don't boil instantly during the day. To envision what it would be like without an atmosphere, we need only travel into space, beyond the bonds of Earth's gravity and onto the surface of our gray, lifeless moon. The moon has an atmosphere, but it is so thin that it may as well be a vacuum. If you were to stand on the bright side of our moon and look at the Earth, you'd quickly fry in temperatures exceeding 100°C. Take a trip to the dark side, and you'd find yourself in temperatures colder than anything on Earth's surface, dropping to a bone-chilling –160°C.

The gases that surround our planet act as a blanket. Like any blanket, the material it is made of determines how warm it keeps us.

The substance we call "air" is actually a soup of gases. By volume, it's 78.09% nitrogen (N_2), 20.95% oxygen (O_2), and about 0.93% argon. The remainder are trace gases and carbon dioxide, or CO_2. CO_2 is a colorless, odorless gas that plays an important role in regulating Earth's temperature. The amount of CO_2 in the air is actually relatively small—as I'm writing this book, CO_2's concentration is just over 400 parts per million (ppm). That means if you were to randomly grab one million molecules of air, 400 of those particles would be carbon dioxide.

The more CO_2 in the atmosphere, the more heat energy is retained at Earth's surface. It basically works like this: energy, in the form of ultraviolet (UV) light, visible light, and near-infrared radiation, is emitted by the sun and bathes the entire planet. As these energy waves shoot through space toward our home, they come into contact with the atmosphere before they can reach the surface. Right away, 26% of these waves bounce right off the atmosphere and head back into space, lost to us forever. The atmosphere itself absorbs 19%, and the remainder make it all the way down to sea level. The ground and oceans then capture some of this energy, and they warm up as a result.

Warm objects emit thermal radiation. Think of it like a stovetop—you don't actually have to touch a stove to feel its heat, because you can sense its warmth by holding your hand a few inches away from it. Earth, heated by the sun, effectively becomes a giant stovetop, radiating heat energy back up into the atmosphere. Some of the heat radiating from the surface goes back out to space, but some of it is absorbed by the air. Once the air itself is warmed, it too must radiate. Some of this energy also travels to space, while the rest goes back down to the planet's surface.

CO_2 is just one of several gases referred to as greenhouse gases, or GHGs. GHGs refer to any type of gas that absorbs and emits thermal radiation. Just like a greenhouse used by a gardener, these gases trap heat energy that would normally leak back out into space. The most abundant gases in our atmosphere (N_2 and O_2) are not GHGs. They

This number, to me, is terrifying. In three decades, we could cross a threshold from which there is no going back within our lifetimes or the lifetimes of our children. If we go beyond that—which the status quo will guarantee that we do, unless major changes are made—then our future could be dark indeed. Kevin Anderson, a climate scientist in the United Kingdom, was quoted in a 2011 article as saying: "A 4-degrees C future is incompatible with an organized global community, is likely to be beyond 'adaptation,' is devastating to the majority of ecosystems, and has a high probability of not being stable."

The clock is ticking. But what is it ticking toward?

THE SCARIEST THING IMAGINABLE

Throughout history, humanity has done a pretty good job at overcoming obstacles. We've had horrific plagues, global-scale war, famines, droughts—but as a species we've always pulled through. We're effective when dealing with a crisis that's right in front of us. Just look at how communities rally when there's a fire, flood, or other natural disaster. People swoop in, often from all over the world, ready to offer money and assistance.

But climate change is different. Aside from potential nuclear war, there hasn't been an existential threat to our species that has as far-reaching implications as climate change. Earth—our home—is being transformed, in a bizarre reverse-terraforming operation that strikes at the most basic life-support systems that have underpinned the development of civilizations over the past ten thousand years. It will hit us on multiple fronts all at the same time, and it's hard to imagine that our ability to adapt will keep pace.

Let's start with the concept of the "average temperature." The mathematical term *average* means you add everything together and divide by the number of items that you added. For example, imagine a hundred cars are driving down a road with a speed limit of 50 km/h. If half the cars were going 40 km/h, and half were going 60 km/h, their average speed would be 50 km/h. Now imagine that the cars' average speed increased. There will still be fast cars and slow cars, but the slower cars will be a little closer to what used to be the average, and the faster ones will be going far beyond the speed limit. Eventually, the cars will be going fast enough that some will begin to crash. Not all the cars will wreck at the same time, but those going the fastest will be most at risk—and when they crash, the effects of that impact will affect every car on the road.

When the average global temperature goes up two degrees, the effects on temperature and weather will be diverse. First, it will be hotter overall. Places that used to average 10°C (50°F) will average 12°C (53.6°F). It doesn't sound like much, but 1 degree represents a big difference, let alone 2. If you don't believe me, try decreasing the temperature of your house by 2 degrees. Where you were comfortable before, you're likely to need a sweater at the new temperature. The second thing that happens is that extreme temperatures become more common. Longer, persistent heat waves increase in frequency. Extreme cold—the likes of which are necessary for maintaining the temperature balance of the planet—will become less cold and will edge closer and closer to the melting point of water. We all know what happens when ice goes from −1°C to +1°C.

The tipping point, where we lose persistent subzero temperatures in cold climates, is probably the most well-known consequence of global warming. Warmer air, particularly at the North and South Poles, means a lot of ice will melt. We're facing the prospect of massive glaciers and ice sheets that have been frozen solid for tens of thousands of years suddenly sliding into the ocean and raising the world's sea levels. We are already observing this now, and with further warming, it will only intensify.

According to a 2013 study in the *Proceedings of the National Academy of Sciences*, for every degree of warming, the seas will probably rise 2 meters (6.6 feet). This is terrible if you're a coastal city. If you're an island nation, it's an existential threat. Nowhere is the story more poignant than on the island nation of Kiribati. This nation, spread over 32 atolls, does not have any land more than 3 m above sea level. For this country of one hundred thousand people, climate change is likely to wipe it off the map. The Maldives, much of Palau, Micronesia, the Cape Verde islands, and several others also risk losing much of their land mass to rising seas. To make matters worse, a melted ocean is also a hotter ocean, and hot water takes up more space than cold water. This is called thermal expansion. So there will be more water in the ocean, the water will have a greater volume, and more and more will continue to flow in from melting glaciers.

All this heat energy in the water adds another dimension to the climate-altered world—the increased frequency of superstorms. Warm ocean temperatures power hurricanes and cyclones. Research has demonstrated that as global warming progresses, storms will become more frequent and more powerful and will make landfall more often than they currently do. So even if you've built your house up on stilts to avoid the sea-level rise, you're likely to get a pounding from hurricanes or cyclones. Superstorm Sandy in New York in 2012 was a chilling example of this. That one storm alone cost $50 billion to clean up.

But let's say you live in a fortress on a hill, well above sea level in a weatherproof bunker. You'll still need water, food, and the relative security that comes with living in an organized society. All of these things will be challenged by climate change.

All the water we drink ultimately comes from the clouds above us, which drop it back to the surface as rain. Even our underground aquifers, from which we get "ground water," depend on precipitation. This fresh water is the lifeblood of every living thing, and climate change will have a big impact on where, when, and how much water falls on the world. Some places will get wetter, others drier, but again the problem is extremes. When a deluge occurs, it will be massive. Conversely, droughts will be longer, more severe, and ultimately more devastating. And droughts exacerbate wildfires, which are becoming an increasingly serious problem, especially in North America and Indonesia.

As the surface warms, mountain ranges will warm up as well. That's bad news for the water supply of any city that depends on snowpack. Over the summer, this snow melts, fueling rivers and streams. As the climate warms, the snowpack vanishes. In a 2015 article in the scientific journal *Environmental Research Letters*, a team of scientists demonstrated that much of the west coast of the United States, as well as vast swaths of Asia and the Middle East, will have their water supplies seriously threatened by the loss of mountain snow. Without reliable sources of fresh water, farming becomes difficult. In 2016, most of the state of California was in its fifth year of a serious drought. There is no sign of relief, and it appears likely that this drought will extend into 2017 and perhaps beyond. Much of the state's agriculture has been threatened, and prices for fruits and vegetables across North America increased as a result.

But maybe you don't think any of this applies to you. Maybe you live in a high place, with greenhouses to grow food and solar panels to generate your own electricity. Maybe you have a rain barrel and other systems to keep yourself hydrated. Even if all these things are true, a climate-stressed world will not be a peaceful one, and that affects everyone, no matter how self-dependent. In a 2003 report, the Pentagon stated, "Disruption and conflict [due to climate change] will be endemic features of life. Once again, warfare would define human life." They refreshed this viewpoint in a 2014 analysis that showed that climate change is one of the biggest threats to the national security of the United States.

For a glimpse into a future of climate warfare, we can look to the present—to Syria, the war-torn nation ripped apart by despots, terrorist groups, and a slew of competing factions. While many factors sparked the conflict and many more keep it inflamed, one of the original problems in the region was a scarcity of water brought on by a major drought that started in 2007. In 2015, a group of scientists demonstrated clearly that this drought was way beyond what you would expect from natural variation alone. This was a climate change drought, and therefore Syria is embroiled in a war exacerbated by a hotter climate. Europe is now dealing with millions of refugees fleeing the conflict, demonstrating that nations far from the war zone will have to deal with the consequences of battle. As the planet's life-support systems fail, so too will the international order that has built a mostly peaceful world. Resource wars, or conflicts based primarily on scarcity of the basic things we need to survive and function, could come to define our lives and those of our children.

This is what I meant when I said that climate change is unlike anything we as a species have overcome before. Even if you survive the extreme temperatures, your city could be drowned by the rising ocean. If you live in a high place, you may get flattened by a major storm. If you weather the storm, you may be faced with thirst and hunger. If you manage to get past that, then be prepared to fend off the climate refugees that weren't so lucky. It's a full-spectrum problem that will only get harder to respond to as conditions deteriorate.

So climate change is bad news for humanity. But what does it mean for the other species with which we share the planet?

THE OCEANS

I'm a marine scientist, so let's start by talking about the oceans. As the oceans warm, the creatures that inhabit them will have to do something to respond to these higher temperatures.

One option is to move. Fish, especially large fish that swim long distances, may migrate in a poleward direction (north, in the Northern Hemisphere, and south in the Southern Hemisphere). If you're a fisherman and the species you normally catch moves outside of your country's fishing grounds—or moves further from your home harbor—then this is not good news. But many species can't move. Fish that are constrained to habitat (for example, fish that live on coral reefs) or stationary species that can't move at all (for example, corals themselves) have no choice but to adapt. Whether adaptation is possible is not always clear, and it's certainly different across species. There are entire branches of research attempting to understand the ability of marine and land-based species to survive warmer conditions, but it's fair to say that they have an uphill battle. It's a hill many species won't be able to climb.

Even if a species can move or adapt to the warmer waters, the chemistry of the ocean itself is changing. I mentioned earlier that much of the CO_2 belched into the atmosphere by our cars, planes, and power plants gets absorbed by the ocean. Through a chemical process, CO_2 reacts with water to form carbonic acid, and as a result the ocean gradually becomes more acidic. Our ocean is a major sponge for carbon, and it soaks up much of the GHGs that we emit every year. In fact, its ability to absorb CO_2 is part of the reason that our climate hasn't warmed faster than it already has. But we are paying dearly for this service.

Acidic oceans are bad for anything that forms a calcium carbonate shell, such as crabs, shrimps, and other crustaceans. We are not expecting these animals to literally dissolve, but it is likely that these species will have to work a lot harder to build and maintain their shells. Extra energy that goes to shell building is energy that doesn't go toward looking for food, reproducing, or any of the other things that marine critters have to do to survive and persist. An ecosystem without crabs, shrimps, and other small-bodied species will eventually collapse, as the species higher on the food chain will have nothing to eat. We're talking about the potential to lose entire ecosystems, and working in concert with other stresses like warmer water, acidity could spell doom for many species.

There is a terrifying similarity between what we are witnessing now and an event called the Permian extinction—basically the worst thing to ever happen to life on Earth. About 252 million years ago, 90% of life in the fossil record simply disappears, in a mass extinction that makes the loss of the dinosaurs look like a pleasant day at the beach. As best as we can tell, this loss was triggered by a massive drop in the ocean's pH, which back then was caused by a massive volume of volcanic gases being emitted into the atmosphere. This acidic environment proved too difficult for much of the life in the ocean. Chillingly, a 2003 paper in the journal *Nature* showed that we stand to produce oceans that are just shy of 10 times more acidic than they are today, and that the ocean will acidify at a rate faster than anything observed in the geological record. It caused a mass extinction last time. What will it do this time?

THE SIXTH MASS EXTINCTION

Most studies agree that we are now in the sixth mass extinction. Unlike the Permian extinction, this one is our fault. For species that live on land, our biggest transgression has been the destruction of their habitat. Development has forced many species to live in smaller and smaller home ranges. Many species on Earth have had their ranges contract, become restricted, or be eliminated altogether by human development. What we don't develop, we often pollute. But both of these issues can be resolved by direct human action. We can set up national parks, and we can make pollution illegal (and enforce the laws and policies).

What we can't do is turn down the temperature when it gets too hot. Once climate disruption has occurred, it is game over for many of the world's imperiled species. Just like humans, all species are constrained in where they can live by temperature. Too cold, and their bodies will freeze, and they will be unable to carry on the metabolic activities necessary to survive. Too hot, and the proteins in their cells break down. These are called "thermal tolerance limits," and species have evolved these limits over millions of years.

Climate change throws these finely tuned relationships out of balance. Again, individuals of a given species have three options: they can move, adapt, or die. If the species is lucky enough to be mobile, it can move in the direction of the poles to seek cooler temperatures. That works just fine, except when the species runs out of room to move—for example, if a North African species is blocked by the Mediterranean, or if a species is constrained by a mountain range. In these cases, their range size simply contracts, pinched up against a wall as their livable habitat gets smaller and smaller. A smaller range means a smaller population, and eventually, the species is gone. When a species has a small range, it's far more vulnerable to random events, such as storms or disease outbreaks, that can hammer the remaining population. This is why forcing species into tiny habitats makes them more vulnerable to extinction, and climate change is the ultimate reducer of habitat size.

We can't rely on adaptation to save ecosystems either. Sure, some species will do better than others, but every species that exists today is a product of millions of years of incremental changes, which come together to make the modern version of each and every type of plant, animal, fungus, bacteria, protist, and archaean on Earth. Every one of these species encompasses volumes of trade-offs made at some point during its evolutionary history. For example, koalas in Australia are specifically adapted to living on eucalyptus trees, eating and metabolizing leaves that are toxic to most other species. This is a huge asset, but it comes at a price. Koalas are obligated to live in these forests and can live nowhere else. If eucalyptus forests are damaged by climate change, development, or any other reason, there's nothing they can do to adapt. Koalas would just stop existing, and that's it.

Polar bears are another example. Evolution has provided them with an exquisite toolkit that enables them to survive in the Arctic. They are incredible hunters and use sea ice to reach prey-rich environments, where they catch, kill, and eat seals that live in the ocean. As sea ice recedes, the bears spend more time on land, away from the food sources that sustain them. They have to move farther to find prey and expend more energy in the process. For a time, it was thought that polar bears could adapt to this—that they could enter a "walking hibernation" as they went from one place to another in search of food. Perhaps unsurprisingly, more recent research has refuted this hypothesis. The Arctic is not a forgiving place, and the polar bears are evolutionarily maxed out. A loss of sea ice and a loss of access to prey will quite simply mean a loss of bears. They can't move away from the new climate, and they can't adapt. The only other option is to die.

It's a story that will be repeated, over and over again, across many species.

TIP OF THE ICEBERG

There is no way that global society can adapt to everything I've outlined so far without incurring a substantial cost. Certain areas will do better than others, but everyone will be affected. But what if it's worse than we think? What if something were to happen that would truly make this an unsolvable problem?

The scariest single aspect of climate change is the risk of positive feedbacks. Roughly speaking, a positive feedback is what happens when a given process produces something that further accelerates that process. It's like a thermostat working in reverse, telling your heater to belt out more warmth the hotter it gets in your house.

Climate change runs the risk of working the same way. Our emission of CO_2 is creating conditions that could trigger the release of other greenhouse gases that have been locked into our planet over millions of years. For example, as Canada's North warms, the permafrost will start to thaw. Permafrost, by definition, is land that normally remains frozen. As it warms, it transitions from a solid surface to a wet bog. Part of this matrix is a vast quantity of GHGs. It is estimated that two trillion tons of CO_2 and billions of tons of methane are locked into the permafrost matrix—and all of it is at risk of release as the polar regions warm. In the ocean, vast undersea methane deposits threaten to be freed from subsea permafrost as the waters warm. We have even less understanding of these.

We are uncovering new problems, new feedbacks, and new threats nearly every day. James Hansen, the former director of NASA and a famous climate scientist, led a 2015 study that showed that as much as 10 feet of sea-level rise could be triggered by positive feedbacks occurring in the Antarctic region. Current research into how climate change will degrade global health is continually being released. New impacts are discovered, and new horrors are constantly presenting themselves.

There's a classic mental image of a cult leader standing at the side of the road waving a cardboard sign that says "The End Is Near!" Denialists sometimes try to invoke this against the scientific community. But scientists are not cult leaders. We analyze data, we study trends, and we revise our hypotheses as more data are collected. Everything that I have presented so far in this chapter is within the bounds of possible outcomes demonstrated by the best available science to date. It could be worse, or it could be a little less bad, and the precise timeline in which everything happens is subject to revision as evidence improves. But if anything, the danger seems to be coming faster than forecasted, and the impacts seem to be more severe.

The most tempting fantasy is to embrace denialism—the idea that this must be all fake, so we needn't bother worrying about it. Denialists have a suite of arguments they deploy to try to muddy people's understanding of the climate problem (and I recommend the website SkepticalScience.com, as it addresses most of them). But perhaps the most misplaced argument is the idea that scientists are just a bunch of liberal propagandists, trained by universities to lie about climate change.

I'm going to let you in on a little secret and tell you about how professional science works. Most of us are motivated to reveal truths about the natural or social worlds and to better understand the way the universe works. However, in terms of how we're assessed as scientists, the only thing that really counts is reputation, as measured by the number of scientific papers we publish and the number of people who talk about our work in their own research papers (we call these "citations"). Our job prospects and career advancement largely boil down to these two metrics. If we're publishing a lot in reputable outlets, and people are talking about our work, we're in good shape.

Doing studies that agree with consensus is no way to build a reputation in science. It's boring. The results are to be expected, and major science journals won't be all that interested in them. By contrast, if you went out and showed convincingly that a well-established theory was completely wrong, you would be instantly famous. If someone could credibly disprove climate change, then that person would immediately win a Nobel Prize and have their work published in the biggest and best journals on the planet. They would be a science superstar and would have money showered upon their research programs by companies and governments that have direct interests in discrediting climate change. They'd be overturning generations of research through their brilliant analysis. If you have gotten this far and you're still skeptical about the reality, cause, or severity of climate change, then I invite you right now to write those doubts into a peer-reviewed paper and get it published. If you do this, you'll get a job for life—a tenured professorship at a prestigious university, because you will have singlehandedly overturned hundreds of years of research. It would be like showing that gravity doesn't exist.

The reality is that this won't happen. The science is sound. More than 99.9% of published research supports the conclusion that climate change is real and caused by human activities. The IPCC, whose documents represent the collective work of hundreds of expert scientists around the world, calculated a 95% certainty that observed warming is our fault. This is unequivocal, at least as far as any scientific model can be. No reputable scientific body on the planet disputes this. If anything, scientists are too conservative. There have been many peer-reviewed articles demonstrating that scientists—including the

IPCC—tend to underrepresent the seriousness of the climate problem, both in terms of the rate at which it's occurring and also in terms of its potential consequences. Scientists want the public to understand that they are objective, and they don't want to instill hopelessness. Therefore, it's often the case that when conclusions are so harrowing that they seem beyond belief, scientists naturally tend to tone the predictions down.

But let me bring this to a more personal level. I'm a conservation biologist who specializes in marine life. I love scuba diving and discovering things in the ocean. I also have a creative streak. In another life I might have been an engineer, designing, building, and creating new things that make our lives a little more interesting or fun. I'm jealous of that discipline—how wonderful it must be to come home at the end of the day and say, "Look what I created!"

Conservation is a frankly devastating field to be in. Much of what we do deals in quantifying how many species are declining or going extinct and in coming up with ways that have the best chance of reducing these problems. We are constantly struggling to get people to pay attention to what we have to say, let alone act on it. We get slandered by the establishment and sometimes even become the victims of physical or legal harassment. In the best case scenario, we might do some work that improves our ability to forecast just how bad the future will be. Maybe some of us will even play a part in saving a species or two.

Those of us in this field have many motivations. Some are compelled by a love for nature, others by a profound compassion for humanity, which depends on nature as well. Most of us live a double life—we are analytical, dispassionate, and logical when doing research but also profoundly loving toward, emotional about, and awed by the beauty of the biosphere. We strive to make the best, most accurate representation of the problem at hand. While there are certainly examples where scientists didn't get something right or even cheated or plagiarized, our discipline is one of the few that is self-correcting. If only denialists could say the same thing.

NO ESCAPE

We live on a planet that is basically a tiny spaceship. If the Earth were the size of a soccer ball, its atmosphere would extend less than a millimeter from the surface of the ball—the width of lead in a mechanical pencil. As far as we know, that thin skin sustains all the life in the universe. Every breath you or any other organism on Earth draws depends on that tiny invisible skin. Everything we do here, stays here.

We cannot escape climate change here on Earth. It is a planetary phenomenon. Some regions will fare better than others—and being from a rich country always helps—but even if your country or city weathers the storm, it could be flooded by conflict and refugees from parts of the world that aren't so lucky. I also feel compelled to point out that we cannot escape by fleeing the planet. We can take baby steps into space, but our ability to colonize another world is simply nonexistent. Even if we were aware of another planet like Earth, with a clean, livable biosphere that could support human life, it will be too far away to reach, even in multiple human lifetimes. As Carl Sagan once put it, this is where we make our stand.

Terraforming a planet like Mars is currently science fiction. The technology currently doesn't exist, even to visit briefly. Besides, if you really think you can just pick up and flee to another world, I suggest you try living in Antarctica first. Compared to Mars, it is a paradise. It has a breathable atmosphere and is surrounded by a bountiful ocean, rich with food and nutrients. But there's a reason it's not populated. Life is far better when you live in an ecosystem with survivable temperatures, rich topsoil, and a pleasant climate that supports a diverse, human-friendly ecosystem. Antarctica doesn't have this, and Mars is incomparably worse.

If escape is not an option then it follows that we need to take aggressive action to prevent the worst effects of climate change. To do so, we're going to have to change our thinking. We often think of conservation as a cost/benefit analysis. When we do this, the questions are often framed in a similar way: How much conservation can we do while still protecting the economy? How much of the ocean should we protect while preserving the livelihoods of fishermen? When is a species so endangered that the costs of bringing it back will be too costly to bear?

When it comes to climate change, we need to get far more serious. If we are being honest in our debates, the questions should be more dramatic: How many countries is it okay to give up to sea-level rise? How many resource wars can we fight and expect to win, and what will their human costs be? How many coastal cities are we happy with losing to chronic superstorms? Do we really need California to exist, or can we let it succumb to permanent drought? These aren't choices we can make lightly, but they are ones we will be faced with if we don't rapidly decarbonize.

By all estimates, the economic cost of climate change will be unprecedented. According to the DARA group and the Climate Vulnerable Forum, two science-focused NGOs (nongovernmental organizations) in Europe, climate change is already costing the global economy $1.2 trillion per year, or about 1.6% of the global GDP. Much of this is due to impacts on food supply and disease in the developing world. By 2030, they estimated that this figure would rise to 3.2% of global GDP. In 2014, the World Health Organization estimated that about one in eight global premature deaths are due to air pollution, and this will only worsen as more people move to polluted environments. As the situation deteriorates, it will hit your personal pocketbook, and the pocketbooks of your children and grandchildren as well. A 2015 study published in *Nature* found that by the year 2100, the income of an average human will be 23 percent lower than it would have been in a world without climate change. Think of the implications of this—trillions of dollars per year and countless lives lost as a result of our damaged climate.

Climate change will cost us dearly no matter what we do. But we are presented with two options. First, we can pay the costs now, and restructure our global economy to carbon neutrality quickly and aggressively. As a result, we would get fresh infrastructure, sustainable energy, and a slew of other benefits. Alternatively, we could decide that this is simply too costly, and do nothing. That just means that we pay later, as we try to pick up the pieces of a war-ravaged, impoverished, and shattered world.

Even if you still think this is all a big hoax or that scientists are exaggerating the risks of the climate problem, I would argue that the pro-climate pathway is better anyway. Rather than buying oil from despotic regimes, we could be producing the energy we need from renewables right here at home. Rather than struggling with deficits caused by volatile oil prices, we could put our people to work in high-quality jobs building a renewable energy grid. Rather than filling air and water with toxic pollutants, our industries could make things cleanly and in a way that keeps us all far healthier than we are today. Instead of destroying old growth forests, we would keep these habitats intact and enjoy the species that live in them for generations to come. Let's plan for the worst and act to avoid it; if we were wrong all along, then the worst thing that could happen is that we'd build a healthier, cleaner, and more socially just world than the one we have today. All of these goals would be desirable even in the complete absence of the climate threat.

While there is no controversy among serious scientists about the gravity of the climate threat, some people feel that focusing so much on carbon means ignoring other, more immediate threats to our planet's life support systems. This viewpoint operates largely on the assumption that climate is too big for us to affect ourselves, but that we can do something about specific environmental problems. In other words, all we can do is act at the local scale, and hope that at the larger scale, collective action will take care of the climate problem. In addition, some environmentalists have criticized the dominance of climate change in the public debate about conservation. After all, we're also in the midst of a biodiversity crisis, and there are smaller-scale tangible actions that can protect species. For example, there are specific air pollutants that are a problem—shouldn't we focus on banning those? And what about water scarcity, plastics, or any other number of problems that we currently face as a society? Why focus so much on carbon, rather than on the myriad of other issues that face the planet?

My response to this line of argument is two-fold.

First, climate change is a conservation issue that affects everything and everyone on Earth. No one is immune from its effects, and everyone contributes and is partly to blame (some of us more than others; we'll cover this in chapter 2). We have to fight it together, and doing so brings a tremendous amount of attention to the concept of protecting the environment as a whole. In 1973, the evolutionary biologist Theodosius Dobzhansky said that nothing in biology makes sense except in the light of evolution. I believe that in conservation, nothing makes sense except in the light of climate change, because its causes and consequences are similarly pervasive. Therefore, I believe that someone who thinks about climate change will more naturally think about other conservation issues as well. It's like the relationship between language and culture. At first, you learn a few words and some basic grammar. As your skills improve, it becomes more comfortable to engage with the people that speak the same language. As a result, you spend more time in that new culture and become more interested in it as well. Climate change works the same way: it's the gateway to broader conservation values, because it makes all conservation personally relevant. As Canadian author Margaret Atwood put it, it's not climate change—it's everything change.

My second argument is that doing something positive for the climate often translates into doing something positive for another aspect of the environment. For example, the beef industry has a terrible carbon footprint (we'll discuss this in chapter 6), and so to fight climate change we need to eat less beef. But that industry is also a disaster for freshwater ecosystems, because it takes a lot of water to raise a cow, and because industrial-scale farms are major sources of water pollution. Here, the climate-friendly action (to eat less beef) is also the action that protects the local environment. Another example would be forestry. A carbon-friendly policy would argue against cutting down old-growth forests. This is entirely complementary with the goal of protecting forest ecosystems.

Conservation is not a zero-sum game. Adding effort to fighting the climate crisis should not take away from conservation action elsewhere, and if it does, then something has gone wrong. Switching from a gas-guzzler to an electric vehicle does not mean that another activist cannot push for bike lanes. We absolutely should be doing all those local-scale things that can protect the environment, while focusing broadly on the climate issue.

So this is the part of the book where we start to turn things around. This is where we turn anger into action. I'm not here to tell you a doom and gloom story, I'm here to get you off the couch and into the fight. Every problem has a solution, and climate change is just like any other problem—but its solutions will be far more complex than anything we've dealt with before.

As we fight climate change, our victories will be measured in ecosystems saved, in emissions avoided, and in a biosphere protected. Whether we achieve this depends on the number and passion of people. In writing this book, I hope to direct your passion in constructive ways and show you how to be a protagonist in your own story against climate change. I will make the case for action and show where we can get the best bang for our buck in our day-to-day lives. I will show how making these decisions doesn't just benefit the planet— it also benefits your own health and bank account as well.

This book is not a complete blueprint for how to build a carbon-neutral society. It does not describe every caveat of every possible way to solve the climate problem. But it operates on the evidence-based understanding that we have to take drastic and immediate action if we hope to solve the climate problem. I will not wrap this book in a blanket of false pragmatism that ignores the seriousness of the issue. I do not acquiesce to people who believe that this is all an overblown hoax. My professional training as a scientist means that I will present things truthfully, but I will not pretend that objectivity requires me to sugarcoat the issue of climate change. It is real, caused by us, and dangerous, and we need to act on it immediately and aggressively. We all need to take action toward eliminating fossil fuels as a primary energy source—that's what the science tells us, and so that's the logic that underpins this book.

The goal of this book is to inspire positive action on climate change in your own life and to help you take actions that reconcile your desire to be a constructive environmentalist with the realities of living in a carbon-intensive society. By laying out a practicable carbon code of conduct, I hope to set the stage for you to personally act on this issue. In chapter 2 I make the argument that you can be an influential agent of change, and that you can truly have a constructive impact on the climate problem. In chapter 3, I argue that we need to have a carbon code of conduct that all of us live by, and I explain how that code can work, and why.

SUMMARY

→ The term **climate change** refers to how our world is getting hotter. It used to be called "global warming," but climate change is a better description, because in some parts of the planet it will get colder, while in others it will get hotter. In some places it will get wetter and in others, drier. But on average, the planet will get hotter, and cascading environmental disaster will follow.

→ There can be no exaggerating the threat posed by climate change. It will cause droughts, famines, wars, and mass extinction of nonhuman species—and we are already seeing its effects. As icecaps melt and the seas rise, it will submerge entire cities and displace millions of people. Agencies ranging from the United Nations to the International Monetary Fund to the Pentagon have stated that climate change is a serious and immediate threat.

→ Climate change is caused by us. **CO$_2$** is the primary gas that drives the problem. There are others as well, such as methane. We refer to these gases collectively as **greenhouse gases**, or **GHGs**. Nearly every technology we use ultimately emits GHGs.

→ Scientists have determined that the world cannot warm more than 2°C on average, or we will experience a planetary calamity. *We are on pace to exceed the global carbon budget in less than three decades.*

→ Halting and reversing climate change requires a massive-scale deployment of clean technologies across the entire planet. But planet-scale action is a product of billions of decisions, made by people just like you.

→ Have hope. *We can fix it, but we must act quickly.*

CHAPTER 2

SOLUTIONS START WITH YOU

Climate change is too big of a problem for one person to make a difference. You're too small to matter. You can't do anything about it anyway. Even if you could, those other people over there are polluting, so you might as well pollute too. And you know what? You don't even have a right to speak, because you consume oil yourself. So shut up.

This is all nonsense of course, but you've probably heard various forms of these arguments from people who try to block progress on climate change. People who oppose conservation have many different motivations, some more benign than others (we'll get into this in chapter 8). But what matters for us is that their arguments ultimately serve as excuses not to act. In this chapter, we're going to tackle some of these excuses head-on and make the case for individual action. You do matter, and I'll show you why.

First of all, I want to talk about you and your personal effect on the climate. You've probably heard of a concept called *carbon footprint*. This is a shorthand way of discussing the volume of greenhouse gases that you, as an individual, are responsible for emitting. People who frequently fly, eat lots of beef, and consume lots of energy have large footprints, while those who live less carbon-intensive lives have small footprints. The term is useful because it acknowledges that not everyone has the same environmental impact. In general, people in rich countries have bigger footprints, and those in poor countries have smaller ones. Climate change is caused by seven billion of these footprints, all adding up to one big impact on our shared atmosphere.

Across the entire planet, the average person's carbon footprint is about 4 tonnes (4.4 US tons) of CO_2-equivalent GHGs per year. However, every country has a different per-capita carbon footprint. In the United States, the average is between 17 and 19 tonnes per year (I'm giving a range because there is some variability in how different agencies calculate these figures). In other words, the typical American has the carbon footprint of more than four average humans on Earth. The implication here is that if you take action as an American, it's akin to four people taking action who live elsewhere in the world. If you're Canadian or Australian, your footprint is probably about the same as an American's and probably a little higher. Some of the worst offenders include the United Arab Emirates (9 times the global average) and Qatar (more than 13 times the global average). So if you're reading this book in a rich country, you are probably causing a lot more than your fair share of carbon pollution.

Depending on your lifestyle, your carbon footprint could be far greater than your country's average. Our footprint is limited only by our money—meaning that if we wanted to, we could produce a staggering amount of emissions, and there is really no policy or stigma against it. If we want to fly every weekend, we are allowed to do so. If we want to eat beef at every meal, there's no rule against it. A rich North American with a carbon-heavy lifestyle could literally account for the footprint of hundreds of people from rural Ethiopia, Bangladesh, or India. We are carbon giants, and so we bear a particular responsibility to fix the problem.

The wealthier you are, the bigger your responsibility to act on climate change. In a 2015 report, the British charity Oxfam calculated that about half of global emissions are produced by the wealthiest 10% of the world. You may be closer to this wealth threshold than you think. According to Credit Suisse, you only need a net worth of $77,000 USD to be included in the global top ten percent. So a large proportion of people in the developed world make up this small sliver of super-consumers, who are directly responsible for causing much of the climate problem. If you think you may fit this description, there's good news—this also means that you possess the ability to have a disproportionately positive impact in the fight against climate change.

I'm going to run a few quotes by you. See if you can guess what they all have in common:

> "Nobody in the world is regulating their oil and gas sector . . . we will not impose unilateral penalties."

> "Another big problem with any Australian emissions reduction scheme is that it would not make a material difference to atmospheric carbon concentrations unless the big international polluters had similar schemes. Australia accounts for about 1 percent of global carbon dioxide emissions."

> "[The United States] account[s] for almost 20 percent of the world's man-made greenhouse emissions . . . We recognize the responsibility to reduce our emissions. We also recognize the other part of the story—that the rest of the world emits 80 percent of all greenhouse gases. And many of those emissions come from developing countries."

The first quote was from Canada's former prime minister Stephen Harper in 2014, regarding regulations for the petroleum sector, including rules about pollution and emissions.[1] The second was from the former prime minister of Australia, Tony Abbott,[2] arguing against improving Australia's carbon footprint. The third statement was by George W. Bush, the former US president, discussing why his administration rejected the Kyoto protocol in 2001 and deflecting at least part of the blame to developing countries.[3]

Each of these arguments shares the same basic implication, dressed up in varying amounts of rhetoric: there's no point in acting unless everyone else acts first. But here, that argument is demonstrably flawed. Collectively, these three world powers directly emit a little under a fifth of the world's GHGs. That is a massive climate impact for only three countries that make up about 5% of the world's population—and yet these leaders pretend that improving their footprint would have no bearing on climate change. Despite their power, they're invoking the same basic argument that your conservative uncle uses at Thanksgiving dinner: Other people aren't conserving, so why should we?

I don't know about you, but my mother always taught me that the bad behavior of others didn't give me an excuse to behave badly myself. For example, if a riot breaks out and you get caught looting, you don't get forgiven simply because everyone else was rioting. If the evidence is there, you'll be charged and prosecuted. If you are caught speeding, you can't expect the officer to let you off the hook because there were other cars going faster than you. Cat-calling and verbal harassment of women is sadly far too common—does this mean that it's okay to be a harasser yourself, because other people are doing it? Of course not. So why do we suddenly accept this argument when we're talking about climate change?

We do good things for each other as well, even if not everyone is equally generous. Many of us donate to charity, despite the fact that not everyone does the same. We hold doors open for people whose hands are full, even though others may not. We shovel the snow off our elderly neighbor's sidewalk just because it's a nice thing to do, even if other people don't show the same courtesy. Progress has never required that everyone move at the exact same pace.

In addition, arguing that we shouldn't conserve because China's footprint is large ignores the fact that China's footprint is at least partially a result of the fact that they manufacture most of the stuff that we buy. So rather than factories in North America and Europe belching out emissions to make our consumer products, we offshore the pollution to China and elsewhere, raising their emissions and making ours look lower by comparison. Remember, we are living in a scenario that is a ticking time bomb, and if that bomb goes off we will all be affected. Humanity has a fixed carbon budget before average global temperatures are forced more than 2°C above pre-industrial levels. Every gram of GHGs we emit between now and then brings us closer to that ceiling, and whether those grams come from smokestacks of Chinese coal plants, from American refineries, or from burping cattle on the Canadian prairies, they cause the greenhouse effect to worsen. We don't have time for weasel arguments—we have to tackle this problem now, as broadly and quickly as we can.

Traditional thought holds that climate change is exacerbated by the "tragedy of the commons," which posits that people will act in their own self-interest when using a shared resource, even if that act harms the group or the resource as a whole. Typically, the individual gains made by spewing out carbon emissions outweigh the collective damage that these emissions cause, at least over the short term. Under this theory, by adopting policies that run counter to climate action, people can "cheat" and get ahead of others who may be acting in ethical ways. The thing is, with new and emerging technologies, the pro-climate decision is usually the decision that is also in one's self-interest. During the industrial revolution it was true that burning more coal was generally associated with having a stronger economy, but that doesn't have to be true with today's technology. It will take a concerted effort to shift our economies away from that model, but the first people, organizations, and nations to do it will be the most prosperous in the future. And while there are many ways to address the tragedy of the commons, economically and otherwise, I don't feel that the existence of this problem is sufficient justification for us to avoid being part of the solution.

Those of us who pollute the most through the choices we make with our collective wealth bear the greatest responsibility for solving the climate problem. To whom much is given, much is expected. Much of the climate solution will come down to people adopting better technologies and different lifestyles. Some of those will be more expensive now but will become cheaper with time. If you have the means, then you have the opportunity to lead the way by using some of these technologies.

As an early adopter, you can have a massive impact. Integrating green technology into your life before it becomes mainstream paves the way for others, who can then afford to follow in your footsteps. This is true at the scale of our communities, but it's also true at the national and international levels. If you can be among the first to do the right thing, it makes it far easier for the next person in line to do the same, both because it helps bring the cost down of whatever conservation action that you take, but also because it makes the climate-friendly technologies and lifestyles normal. We discuss this further in part II.

I HOPE THEY CYCLED TO THAT PROTEST

One of the least sophisticated ways that people justify inaction on climate change is when they attack activists for participating in the carbon economy. This is the classic complaint that Al Gore shouldn't talk about climate change because he flies in planes, or that activists shouldn't protest petroleum because they drove their car to the rally.

First, it is important to understand that it is entirely legitimate to work within a system to change it. Western society, which is based on participatory democracy, is fundamentally built on this principle. For example, let's imagine a person who is completely opposed to taxes. They stage protests, they organize a movement, and they endorse candidates for office who pledge to reduce taxes. They are a single-issue voter, and that issue is giving less money to government.

But that person exists in a society showered in government services. Roads, libraries, hospitals—all were built using tax dollars.

Whether you agree with our hypothetical fiscal conservative or not, I can't imagine anyone seriously arguing that this person doesn't have a right to voice that opinion because they drive on roads or because they check out books at the library. The apparent hypocrisy may be noted, but this activist would not be silenced by media or public opinion. Disagreed with, maybe, but not silenced.

But this is the exact argument that many people make when they say that climate activists can't speak if they ever used oil. It's nonsense, and it's a catch-22. If activists did swear off fossil fuels entirely, then they would effectively have to separate entirely from modern life. That would rob them of their ability to improve the system. It's hard to run a movement from an off-grid log cabin. But that's exactly what anti-conservationists want—to force people with new ideas to the margins, where they can't push for change.

When it comes to doing good in a fundamentally flawed world, where can we look for inspiration? Despite this being a science-based book, religion holds some lessons for us that could offer some guidance. According to Christian doctrine, we are all sinners. When Adam and Eve were booted out of the Garden of Eden, humanity would be forever tainted by sin. From that point on, the best any of us could accomplish was to sin as little as possible—we are fundamentally sinners, and our job as people with free will is to live according to sets of rules and guidelines laid out in various teachings. As far as I can tell, there is no way to actually live completely free of sin, nor is it really expected.

But that doesn't give you carte blanche to go out and sin as much as you want. You're not supposed to take the concept of original sin and use it as an excuse to engage in all kinds of mayhem. On the contrary, you're supposed to live a life as close as possible to Christianity's code of conduct, as laid out in the Bible. Just because you've failed to not covet doesn't mean that you get to steal. What you're supposed to do is confess your sins (if you're Catholic, anyway) and move on, with the goal being to do better next time.

Carbon emissions work in a similar way. We are all polluters, just as we are all sinners. It is within our power to pollute as little as possible and to do our best as individuals to reduce the amount of pollution that exists in the world. We will never be free of it, but we can always strive to do better.

So we all bear a moral responsibility to moderate our own use of carbon, and that responsibility is larger the more wealth we possess. We can directly reduce our own consumption, but perhaps an even bigger impact of this is that our good behaviors can lead to others changing their behaviors as well. To fight climate change effectively, we have to inspire others to make changes and make conservation "go viral." We need to engage in conspicuous conservation.

Conspicuous conservation means engaging in pro-climate activities that are visible and that make others want to follow suit. We need to loudly, proudly reduce our carbon footprint, and do it in a way that makes others want to emulate us. If you do it, other people will get interested. If enough people do it, it will become the normal way of doing things. Living happily with a low-carbon lifestyle demonstrates that decarbonization is possible and puts the lie to the idea that having a low-carbon footprint means living a dull, uninspiring, or poverty-stricken life.

This is exactly what we're starting to see happen in the world today. Many nations are embracing low-carbon development as a way to bring safe, cheap, and reliable electricity to their populations. Countries as diverse as Morocco, Chile, Ethiopia, and South Africa are aggressively developing renewable energy, choosing the path of wind and solar rather than coal and oil. In doing this, they are pushing back against the idea that green development is something solely for the super wealthy. With the overwhelming majority of the world's energy consumption still coming from fossil fuels, it is not yet mainstream to forgo building a coal power plant. Yet every day we're seeing signs of hope.

The shift is happening politically as well. Across the world, for the first time members of national "green parties" are being elected to office, despite the many barriers to entry for new political entities. Their primary mission is environmental, and climate change is a big part of their agenda. With 5–6% of the entire populations of countries voting on this singular issue, surely it signifies a shift in thinking. And this shift is reflected in the platforms of many larger political parties as well, who are starting to include explicit language and goals around climate change in their national development plans.

Helping conservation go viral is one of the most important things an individual can do, and it starts with living a climate-friendly lifestyle. This is where you, as an individual, can make a tremendous contribution to the global fight against climate change. Conservationists are often outspoken about how governments, corporations, and so on need to make better climate decisions, and this is absolutely true. But they will be more likely to pay attention if our actions match our words. They are the suppliers, and we represent demand. As an individual, you can do your part to express your demands that suppliers abide by carbon-friendly policies and contribute to alleviating climate change. If it becomes impossible to win an election while denying climate change or to earn a profit while promoting unsustainable development, then the actions of these large polluters will be forced to change.

A 2016 study in the science journal *Climatic Change* found strong support for the idea that we need to match our actions to our words when trying to convince others of the merits of fighting climate change. The authors of this study conducted a survey of over two thousand people across the political spectrum, interviewing them about their views on the credibility of a hypothetical climate researcher giving a talk about climate change. The message was clear; when this hypothetical researcher was described as having a large personal carbon footprint, the public was unlikely to accept the importance and validity of the researcher's argument that we all need to reduce our impacts on the atmosphere. By contrast, when the researcher was described as having a low personal footprint, they were seen as far more credible. While the results make intuitive sense, this piece of research was a groundbreaking demonstration of how important it is for us to practice what we preach in the area of climate advocacy.

THE SELFISH CONSERVATIONIST

For many people, changing the world is a little too overwhelming to be a personal goal. Just getting by can be hard enough as it is. I perfectly understand that. We all have lives to lead and our own problems to overcome. However, with the speed at which the climate is changing and the urgency growing for governments to take action, it is in your personal best interest to get ahead of the problem and get set up for a low-carbon lifestyle. Acting in a pro-climate fashion is usually directly in your best interest, or in the interest of the city, province/state, or country in which you live—and if it isn't now, then it will be in the near future.

In a climate-disrupted world, you will still be better off living in an energy-efficient house. In a planet plagued by drought and food shortages, you may as well get used to a low-meat diet. If fossil fuels become scarce or rationed, then a few solar panels could make all the difference in your quality of life—and an electric vehicle would be far better than one that requires petroleum to run. So making all the changes that will benefit your carbon footprint will also make the climate-altered future more survivable for you and your family.

This brings me to my final argument for this chapter, as to why you, as an individual, matter in the struggle against climate change.

In December 2015, the world's governments got together in Paris, France, in the 21st Conference of Parties to the United Nations Framework Convention on Climate Change. There, the world's governments agreed to hold warming to "well below 2 degrees Celsius, while urging efforts to limit the increase to 1.5 degrees." Assuming it is ratified by enough parties, this is a big deal. Whole economies will have to fundamentally shift to achieve this target. But there is a problem.

When you add up the pledges that each country made to reduce emissions, you still get almost 3 degrees of warming. Despite setting ambitious targets, our leaders did not agree to take sufficient action to achieve these targets. Left to their own devices, the governments will not get us where we need to be. Governments can coordinate, and they can deploy resources, but ultimately climate change is the result of the collective actions of each and every one of us. We're all pouring carbon into the atmosphere, and so solving the problem means getting every person to reduce the amount that they produce—starting with those of us who produce the most.

There are champions in this effort. Former president Barack Obama and his administration made concrete steps to improve the carbon footprint of the US economy. Many prominent US governors, including Arnold Schwarzenegger and Jerry Brown, have been vocal supporters of climate action. Bill Gates has promised to spend $2 billion of his own money on clean energy. Tesla CEO Elon Musk has devoted much of his life's work to building the tools for a clean energy future. Pope Francis has been very clear—climate change is a moral issue, and we all need to play a part. UN Secretary General Ban Ki-moon has said repeatedly that climate change represents a serious barrier to the eradication of poverty. And there are armies of artists, musicians, celebrities, and other cultural figures that support climate action as well. You are not alone.

I argued in chapter 1 that climate change is an existential threat to a functioning, civilized global society. Although they probably wouldn't admit it, the fact that most of the world's governments agreed to take action on this problem suggests that they're scared, too.

There aren't many examples we can draw from of global-scale threats that were this serious. World War II is one exception. In that conflict, there was a very real chance of the world being completely taken by evil regimes. Between them, the Axis powers occupied basically all of Europe, and they held a massive sphere of control in the

Pacific. They slaughtered civilians and military alike and enacted the Holocaust. At the time, it must have seemed hopeless—like the end of the world had come.

In response to this existential threat, the Allied nations invoked a doctrine of total war. Total war is the condition where conflict envelopes all aspects of a society. In this situation entire countries built their industrial engines around winning the war. Car factories became tank factories. A single Ford Motor Company plant was building a B-24 bomber every 59 minutes. Universities designed weapons and analyzed statistics relevant to winning battles. Every aspect of society became focused on the goal of winning the conflict. No one was exempt from the effort. Civilians in noncombat roles assisted in the conflict directly, by producing labor that helped produce tools and weapons for the fight. They also assisted indirectly, by rationing, recycling, or investing in war bonds.

This is the mindset we need to apply to solving climate change. No one is exempt from its effects, so no one is exempt from working to fix it—and those of us with the means have a greater responsibility to do so. The great economies of the world will have to restructure, and the language we use to describe the problem will have to change. We will have to retire a whole generation of technology that is dependent on fossil fuels. The task is Herculean. But we, as individuals, can help to push this change.

This is especially true for those of us who hold environmental values. If we aren't leaders here, then who will be?

The most important thing to take away from this chapter is that if we allow ourselves to wallow in the nihilistic belief that climate change is unsolvable, then we are similar to the people who make it their life's work to prevent measures to heal our biosphere. The belief that all we should do is give up, retreat, and surrender the planet can do just as much to prevent real action as any coal baron or oil lobbyist. Denying the hopefulness of the mission is engaging in our own type of selfish climate denial—the denial that immediate, aggressive action has any hope of benefitting our and future generations. It can, and it will.

Even if we can't stay under the 2-degree limit, there are still varying levels of calamity. The hotter it gets, the worse off we'll all be. We can stay on the lower end of that range of possible outcomes if we act right now. It is probably inevitable that we are going to lose many, many species to climate change. It may be inevitable that cities are lost to the sea. There's no getting around the prospect of droughts and resource wars, and the creation of climate refugees is certain. But there's a big difference between a world afflicted by 2-degree warming and one warmed by 3, 4, or even more degrees. Each tenth of a degree we march up the scale will cost lives, species, and money, and the cost will get exponentially higher the further we move up the heat curve.

When it comes to assessing the odds of successfully resolving the climate crisis, I am neither an optimist nor a pessimist. As a scientist, I try to reject both slants and form an objective opinion about the situation. Here's what I believe: There are many talented expert scientists, economists, and policy analysts who have demonstrated clearly what needs to be done to forestall climate disaster. An equally heroic army of engineers is designing the clean, renewable technology that we need to execute this plan. And dedicated NGOs, environmentalists, social justice activists, indigenous groups, and a vast array of other sectors are applying the pressure and coordination needed to make these changes happen.

The path between where we are now and a sustainable future will be made up of billions of small decisions that each person on Earth will have to make. Every one of these decisions will affect our shared outcome. I believe that if you become a champion for good decisions and get your friends and colleagues to do the same, we can prevent the worst parts of a climate-altered future and weather the aspects that are unavoidable.

This book introduces a carbon code of conduct because we need to lay out rules that we should follow to guide these decisions. The carbon code is a path forward for individuals to contribute positively to the battle against climate change, and it gives a structure to follow as you're embarking on this journey. With the carbon code in hand, you will be ready to take action in the best ways that you can, as quickly and aggressively as you are able to. Critically, you will do all these things without sacrificing your ability to take care of yourself and your family.

The carbon code doesn't ask that you walk away from modern life, nor does it ask that you take extraordinary steps that put you far outside the mainstream of society. It does ask that you take a long, hard look at your own behaviors and seek ways to be a better environmental champion. In almost every case, the clean, sustainable option that has a lower carbon footprint is also the choice that will lead to a better quality of life. You will be happier, healthier, and wealthier in a world where you're battling climate change.

I believe that it is not hopeless. In fact, I feel that we are at the start of something amazing. It's not too late to preserve the integrity of our planet's life-support systems, if we start acting right here, right now.

SUMMARY

→ One of the most dangerous lies that climate change denialists deploy is that climate change is too big to fix—and that you, as an individual, are too small to matter.

→ **You absolutely matter**—and everything you do to fight climate change matters too.

→ *Only you* are responsible for your carbon footprint. Achieving a low-carbon lifestyle is easier in some places than others, but ultimately *your decisions determine your footprint.*

→ If you are a serious advocate for climate action, *you must demonstrate this action yourself.* Climate change is a product of collective decisions and behaviors of people all over the world. If you live in a rich country, you are probably emitting much more than your fair share of carbon pollution. *Solving climate change starts with you.*

➤ When climate villains say your actions don't matter, *they are wrong.* When they say that the actions of their business or government are too small to make a difference, *they are incorrect.* Every bit of GHGs emitted brings us closer to global catastrophe, and consequently every piece of GHGs *not* emitted is a concrete step in the right direction.

➤ *You can be the hero of your own climate change battle. Or you can be part of the problem. The choice is yours—and your actions matter.*

CHAPTER 3

THE CARBON
CODE OF CONDUCT

O ur climate is taking a beating, and the planet is getting hotter. It's happening because of us—because we're emitting greenhouse gases into the atmosphere at a rate beyond what the planet can sustain. If it continues, we're all in big trouble. And yet, despite the fact that the vast majority of people accept the reality of climate change, there is no real stigma against burning fossil fuels or emitting CO_2.

Let's perform a thought experiment. Imagine that you had a big barrel of gasoline. Imagine taking that barrel, rolling it over to the nearest storm drain, and tipping it over in front of a crowd of onlookers. Imagine liters and liters of petroleum flowing down into the storm drain, making its way to an urban stream. Now imagine looking at the expressions of the onlookers. What do you think their reaction would be? What would the online comments section look like if this scenario were filmed and uploaded? I bet it wouldn't be positive.

People do care about the environment, and this is a prime example. But let's change our thought experiment. What if, instead of dumping a barrel of gasoline, you poured it into a vehicle's gas tank and burned it? Rather than spilling fuel into your urban stream, you're spilling it into the airshed. How would people react in this case? We both know the answer to that. When GHGs "disappear" into the atmosphere, we just don't make the mental link between emission and carbon pollution. Our societal norm is that invisible carbon emissions are mostly okay, even if people consider themselves to be concerned with climate change.

We don't treat all air pollution with such ambivalence. For example, after scientists demonstrated that sulfur dioxide caused acid rain, the federal government of the United States passed the Clean Air Act to greatly restrict these emissions. The public pushed the government to act, based on some very good science done to show the extent and severity of the problem. Society collectively decided that these unregulated emissions could no longer be considered normal, and as a result the government was pressured to do something about it. Smoking is another example. You don't have to look too far into the past to find restaurants and bars that were hazy with cigarette smoke. Nowadays, smoking in public places is often illegal, and it would be a major faux pas to walk into someone's home and light up without permission.

When it comes to carbon emissions, it's difficult to agree on what should be considered "normal." We are currently dependent on technologies that emit GHGs, and since everyone uses them, how can there be anything wrong with them? Many environmentally minded folks drive themselves to work, fly around the world, and purchase electricity from coal-fired power plants. In some cases, we see alternatives as expensive, impractical, or nonexistent. But often we don't think about it at all. At most, we may feel bad about our personal footprint, but there is no real social pressure to reduce it, even though on aggregate the consequences of these emissions are staggering.

One of my favorite environmental ad campaigns was conducted by the electrical company in British Columbia, Canada.[1] The ad showed people going about their daily routine, eating apples, shaving, or packing lunches in aluminum foil. In each case, the person was being hilariously wasteful—taking a bite out of one apple and throwing it on the ground, making one stroke of the razor blade and disposing of it, and using an entire roll of foil for a sandwich. The ad then switched to a light bulb that was left on all day with no one at home, wasting electricity. The punchline: "The most ridiculous thing about wasting power is that, for some reason, we don't think it's ridiculous."

In chapter 2, we discussed why your pro-climate actions matter, and why it is legitimate to work within a system to change the way the system works (i.e., you don't have to forego fossil fuels entirely to be a positive force). We can't stand idly by while big companies and governments act in ways that harm the climate. But we also can't let ourselves off the hook. The accusation of hypocrisy that people deploy against those who call for change is a powerful one, and it is something that we need to address directly. Sometimes, these folks have a point. Think of the environmentalists whose jet-setting habits give them a carbon footprint far higher than the average global citizens'. Or the marine conservationist who regularly eats endangered fish. We have to do better—holding the right beliefs is not enough. Science has shown that we are best able to win hearts and minds if we live the life that we espouse and demonstrate that it's actually a pretty great way to exist.

Where do we draw the line between someone who is wasting carbon and someone who is using it responsibly? How can we judge our own behaviors to assess whether we're using carbon ethically?

The answer is to develop and live by a carbon code of conduct.

BUILDING THE CODE

A code of conduct is a set of rules that outline norms and expectations. We use them where behaviors need to be guided and structured. A good code of conduct will include principles that a group expects its members to follow, and it should be posted publicly. This accomplishes two things. First, it lets everyone in the group know what is expected of them. Second, it shows everyone outside of the group how people within that organization are supposed to be behaving. Codes of conduct are not the same thing as laws—laws are dichotomous, and you are either following them or you're not. Codes of conduct are interpretable and general and are designed to shape expectations. They provide clarity where things were previously ambiguous—they're designed for the "gray areas" of our lives, where the right thing to do is not always obvious.

We see codes of conduct all over the place. Google's parent company, Alphabet, has one, and its primary motto is "Do the right thing." You can find it on their website. Amazon, Apple, Dell, and a litany of other major organizations have employee codes of conduct that establish their culture and expectations. Medical doctors have a modern code of ethics based on the Hippocratic Oath. Nearly every professional society bears a code of ethics or code of conduct that governs the practices of its members.

My favorite code of conduct belonged to the starship *Enterprise* from the television show *Star Trek*. Its premise was so wonderfully simple: *To explore strange new worlds. To seek out new life and new civilizations. To boldly go where no one has gone before.*

What should we do? Explore strange new worlds.

Why are we doing it? To find new life and new civilizations.

How should we do it? Boldly. It's all there, and it's all clear.

For carbon use, we are in desperate need of a code of conduct, to provide guidance on how to reconcile our concerns about climate change with our need to emit carbon to participate in modern society. This is especially essential for those of us who emit a lot of carbon ourselves in the service of the greater mission of carbon neutrality.

Our code needs to meet a few criteria if it's going to help us solve the climate problem. First, it needs to be scalable. Climate change is a difficult problem because it can only be fixed by collective action—yet collective action is merely a sum of a whole lot of individual actions. So it has to be something you can use within your own life that can be scaled up for use within a company, government, or other organization that can guide the action of its members and its primary activities. Second, the code should be flexible. Everyone contributes to carbon pollution, but some contribute more than others. Therefore we need a set of rules that we can apply whether the organization's footprint is big or small. For example, the carbon code of conduct for an airline would look quite different from that of a corner store. Both matter, but the former stresses the climate in a different way than the latter. Third, it needs to be simple and clear. The best mottoes and codes of conduct are so succinct that they burn themselves into your memory. The basic principles of a carbon code should be equally memorable. Fourth, it should help us avoid information paralysis—the problem that occurs when we fail to act because we think we don't fully understand the problem. We need to know enough to act in ways that benefit the climate, but it is certainly possible to do the right thing most of the time without having a complete understanding of every aspect of our own footprint. Finally, the code should be effective at guiding people's decisions, so that we can make a measurable dent in our footprints or the footprints of the organizations to which we belong.

A good code of conduct should help us reduce the ambiguity in what is considered ethical for carbon use. For instance, when is it okay for a conservation biologist to fly across the world for a scientific conference? When is it okay to drive a gas-guzzling four-by-four vehicle? How can we lead happy fulfilling lives while limiting our use of carbon—and specifically, how can we do this without putting ourselves, as individuals, at risk of falling behind the rest of the pack?

The carbon code of conduct consists of four "R" principles that must be applied to one's carbon usage:

1. **Reduce** *your use of carbon as much as possible.*
2. **Replace** *carbon-intensive activities with those that use less carbon to achieve the same outcome.*
3. **Refine** *the activity to get the most benefit for each unit of carbon emitted.*
4. *Finally,* **Rehabilitate** *the atmosphere by offsetting carbon usage.*

If you're burning fossil fuels expressly in support of conservation (e.g., if you're conducting scientific research, attending a climate change conference, etc.), then one further question should guide your decision: If I use this carbon, is there a reasonable chance that the activity will contribute to a net reduction in humanity's carbon footprint or a net increase in the biosphere's resilience to climate change? If you can't answer yes, then you need to cycle back to the first "R" and look for ways to reduce the footprint of your activity. This is particularly important for those of us who aim to serve as role models for the rest of society by being conspicuously ethical in our emission of GHGs.

These rules, taken together, form the basis of the carbon code of conduct. They are simple but profound. They are scalable—they work on the individual level, for your day-to-day decisions, but also make sense when applied to large groups and organizations.

Now, I didn't pull these rules out of thin air. In fact, I'm merely repurposing rules that the scientific community has already agreed on after decades of debate and study on a similar issue.

DOING HARM IN THE NAME OF PROGRESS

Why is the carbon code structured around these four Rs? To understand its logic, we have to take a step back and think about why it is that we emit carbon in the first place. When we use fossil fuels, we do so because it's practically impossible to avoid them completely and still participate in modern society. In our current system, emitting no GHGs at all would mean that we are not involving ourselves in the carbon-based economy and as a result will have trouble promoting pro-climate behaviors, policies, and technologies. For conservationists using GHGs toward professional ends, the goal must be to strive for a net reduction in GHG emissions as a result of their carbon-using activity. For example, we may need to fly to a conference on climate change, which could play a role in enacting pro-climate policies. Or we may operate a mine that harvests rare earth metals—critical components in renewable energy technology.

In other words, when we make a decision about using carbon, we have to balance harm with effectiveness. This is not a novel concept.

In fact, the scientific community has considered this issue for a long time.

Since the dawn of medical research, scientists have fielded questions about the need to harm animals in the process of testing treatments that can save the lives of humans. The controversy around this practice can be summed up as follows: it is unethical to cause harm to animals, yet many scientists feel we must do so to advance the science and practice of medicine. The debate on how to reconcile the need to advance medical knowledge with society's growing demand to treat animals ethically has raged for as long as scientists have conducted biological research.

For many years, dogs, pigs, monkeys, and other species were experimented on with few rules or standards. Animal care varied from study to study, and conditions were often deplorable. Animals were subjected to horrific procedures with no regard for pain and suffering and were kept alive far beyond the point where useful data were being collected. Since at least the early 1800s, people inside and outside the scientific community became vocally concerned about the unrestricted use of animals in research and the pain and suffering that these studies were causing. Despite the growing concern, the majority view was that there was no good reason to change (sound familiar?).

Scientists have a lot of good reasons to test things on animals. For one, most people believe that testing serves a greater good—it would be more unethical to test procedures directly on humans or even to not test at all and allow diseases to go untreated. Second, there is often no good alternative. Biological processes are complicated, and to determine whether a treatment would work on patients, things have to be tested on an organism that is as similar to humans as possible. Third, scientists have to use substantial numbers of animals because not every animal reacts the same way to a treatment, and so medical science is based on looking at the average effect of treatment on a large number of animals.

While all these facts are true, it is also true that there needed to be rules in place to govern the practice of animal use in science. To retain its social license, science had to change. The modern guidelines for animal experimentation were outlined in a book published in 1959, entitled *The Principles of Humane Experimental Technique*, by William Russell and Rex Burch. In this book, the authors conducted a thorough review of the ethics of using animals for research purposes and concluded that while animal testing remained necessary, scientists had to agree to a set of standards of practice to govern when and to what degree animal testing is acceptable. They had to develop a code of conduct.

It is their code that I have adopted to address our modern-day carbon problem. The principles that underpinned their code, like the carbon code above, were reduction, replacement, and refinement.

In the context of animal testing, reduction means that researchers are obligated to use the minimum number of animals necessary to complete a study. Scientists must estimate in advance the lowest number of animals needed to draw meaningful conclusions from their study. Too many, and the study will have caused unnecessary pain and suffering, as the same benefits could have been achieved by harming fewer animals. Too few, and the animals used in the study will have suffered in vain, as the scientists would be unable to tell whether their results were real or just due to chance alone.

Replacement refers to substituting highly sentient animals with those that are less sentient or with different methodologies that eliminate the need to test on animals at all. An example would be replacing chimpanzees with lab mice. Better still would be to study invertebrates, which probably don't feel pain the same way we do. However, this can be risky—deviate too far from human physiology, and your results may become useless. Therefore, replacement requires a balancing act between the sentience of the species on which one experiments and the need to get usable data that transfers to humans. Replacement can also mean substituting animal-based science with computer simulations. This is sometimes done when training medical students in surgical procedures, though the extent to which it is suitable is debated.

Refinement in animal research focuses primarily on improving the conditions experienced by animals in the study. This means selecting methods that are less invasive or that alleviate pain and distress. In effect, the principle of refinement means that we must do everything we can to reduce the suffering of the animal, provided that the intervention does not reduce the quality of data gathered in the study. For example, we might provide mice with shelters or comfortable surfaces to rest on, rather than a bare metal cage. Refinement also requires that we formally determine the end point of the study— that is, the point at which the animal is rehabilitated and released or euthanized.

These principles represent the three Rs of animal care, agreed upon by most scientists for many decades. Rehabilitation was added more recently by some science bodies. This principle states that if an animal is not euthanized in the study, then scientists are obligated to nurse it back to health.

Taken together, these Rs have had a profound effect on biomedical research. They provide a common set of expectations for how scientists use nonhuman species. In fact, this code of animal use has been formally enshrined in the laws of some countries. The existence of this code has not ground scientific progress to a halt. By contrast, having an agreed-upon set of rules for animal research has given scientists the social license they need to conduct their work. The Rs of animal testing govern scientific research to this day, and while views on what constitutes ethical treatment of animals continue to evolve, the Rs themselves—and the requirements they place on scientists—have stood the test of time. This code remains the guiding principle against which all biological research is tested, and without having approval under these rules, the study cannot proceed. I am held to these standards in my own ecological and conservation research, and I am grateful for them. In following this code of conduct, I know that I can sleep well at night—I have done my job ethically and against a standard that is agreed upon by science and most of society.

What does animal testing have to do with carbon emissions? For one, it is unethical to harm animals, just as it is unethical to doom human and nonhuman species to the impacts of climate change by emitting carbon—yet scientists must do the former to advance biological knowledge, and we all must do some of the latter to exist as productive members of society. Secondly, techniques are continually advancing that can empower us to do less harm and get more benefit per unit harm done. Third, it is difficult to imagine a world completely absent of either animal use or carbon emissions (for now, anyway), so we must continue to balance harm with progress. Fourth, in addition to the direct harm caused by the action, the action itself can disempower the actor to cause positive change. In other words, someone who causes undue harm to the climate in the name of conservation, without considering alternatives, will be harder to take seriously than someone who practices what they preach.

Completing this analogy, I believe that we can learn lessons from the myriad of debates that ultimately resulted in a shared code of conduct for animal usage. We don't need to spend decades, as animal researchers did, trying to figure out the framework for a carbon code. The answer lies in agreeing upon a new norm for carbon emissions based on the four Rs. These principles could guide the way we use carbon in an environmentally minded society. That is why we should group these guidelines together into a carbon code of conduct.

STREAMLINING DECISIONS

A carbon code based on reduction, replacement, refinement, and rehabilitation will make it easier to be a good advocate for the atmosphere. It will help us streamline decision making in our personal lives and structure decision making in the organizations in which we participate. But making decisions like these, in isolation, can be a pretty exhausting process. One of the benefits of having a carbon code is that it nests all your little decisions into one big decision to be a constructive advocate for the atmosphere.

To explain what I mean, I'd like to take you back to my life before science, when I was a competitive swimmer in my home town of Burnaby, British Columbia. I made it pretty far in the swimming world and capped off my athletic career as a member of Simon Fraser University's varsity swim team. I was never a natural athlete, and every step I took in the sport was rugged and hard fought. I slowly climbed up the ranks of the sport, always starting from the bottom of the pack. My wins were few and far between but were extremely gratifying and hard earned. I advanced with sheer grit and determination. I was envious of the people who made it look easy but plugged along all the same.

The lifestyle of a swimmer is very intense. We trained—a lot. We were in the water twice a day during the week and every Saturday morning. The alarm clock rang at 4 a.m., and we were in the pool by 5:30. Every time I dove into the water I felt like I was in a fight for my life. Each practice was a race, and we all suffered. If we had a bad day, we just suffered more—there was no way out of a swim practice. It bred toughness, but it was also incredibly taxing, both mentally and physically. Every day I had to make a decision to subject myself to more pain and suffering in pursuit of a greater goal. Think of how hard it is to decide to go to the gym after a long day at work—swimmers and many other athletes make this decision every day, and the answer always has to be yes.

The attitude behind this is described very well in a 1998 book written by Keith Bell, a sports psychologist, called the *Swim to Win Playbook*. The major point of this book was that deciding is hard, so in order to win you have to make a single big decision that guides everything you do. For example, your goal might be to win an Olympic gold medal. If that is your goal, then you commit to a grand decision to achieve

Olympic glory. As a result, all your other decisions that support this large decision are made automatically.

For example, "I'm tired—should I go to practice?" is met with a resounding internal response of "Of course I should." "Should I grab dinner at the bar with my friends, or should I go home and make something healthy?" Again, the decision is premade. In this model, the big commitment streamlines your mind so that you can achieve your goals.

Something akin to this model was referenced by US President Barack Obama, in a 2012 interview with *Vanity Fair*:[2] "You'll see I wear only gray or blue suits . . . I'm trying to pare down decisions. I don't want to make decisions about what I'm eating or wearing. Because I have too many other decisions to make . . . You need to focus your decision-making energy. You need to routinize yourself. You can't be going through the day distracted by trivia."

The concept here is similar—make a big commitment to abide by the carbon code of conduct, which streamlines your decisions around your use of carbon. Should I reduce, replace, or refine? Yes, because I have committed to doing my part to solve the climate crisis. I have committed to a carbon code of conduct.

If it works for professional athletes, and it works for a president of the United States, then I think it can work here. Every day we make a litany of little decisions—do I walk, take the bus, or drive? Do I buy a bunch of beef for dinner or eat chicken or fish instead? Do I bring reusable bags to the store, or do I take plastic? Each of these decisions has a consequence, and each one matters. Each decision stands to either reduce or increase our carbon burden, and each one can be an opportunity to engage in conspicuous conservation in support of the larger goal.

Committing to following a carbon code of conduct is our big decision. It means that whenever we do something that uses fossil fuels, we quickly run through the four Rs. Do I really need to do this, or can I reduce the activity in some way? Can I replace this activity with something less impactful? How do I refine it to get more benefit out of the carbon I use? And if all else fails, is there an appropriate offset, so that I can rehabilitate the climate for the carbon that I use?

The carbon code is intellectual shorthand—a quick checklist I run through over and over when I make the little decisions that make up my daily life. Like any habit, you just have to practice it regularly, and it becomes automatic. As I've written this book, I've found that the carbon code has been very easy to apply in practice, and it has made me a far more responsible conservationist. It has not gotten in the way—on the contrary, I am now more efficient with my time and resources, thanks to the principles laid out within it.

Having a carbon code of conduct provides a way for you to live your life to the fullest, while being a clear positive force on the climate. It is not a call for a purely minimalist lifestyle. I am a very practical person. I like technology, and I like living in a modern world. I think a lot of people like it—and there are many who don't live like the average North American but wish they could. I don't think it is realistic to expect people, on their own, to give up their place in society and take a vow of poverty. However, as we made clear in chapter 1, there are no bystanders in this fight. So think of the carbon code as your guidebook for the struggle.

When you're ready to do so, I'd like you to write a personal commitment to fighting climate change. This will be your personal climate commitment. It needs to be clear and include two parts—a commitment to reducing your own carbon footprint as much as possible, and a commitment to convincing others to do the same.

If you'd like, I've provided a sample climate commitment you can sign below:

Date: _____

I, _____, am making a personal commitment to solving climate change. I commit to applying the carbon code of conduct to my daily life and will **reduce**, **replace**, **refine**, and

rehabilitate my use of carbon. I commit to convincing others to follow this code as well. I do this because of my love for the biosphere, my love for humanity, and my desire to live a healthy and sustainable life.

At this point in the book, I've explained the dangers of climate change and have argued that it is the major struggle that our generation has to face. I've argued that each and every one of us has to take conspicuous action, to the extent that we are able, and the richer we are the more obligation we have to act. With the carbon code, we now have the simple principles that can help us reconcile our use of carbon with our mission to build a sustainable future.

Learning how to apply those principles is the focus of part II. We'll review the activities that make up the biggest component of most people's carbon footprints in the developed world and identify ways to reduce these footprints through the four Rs of the carbon code. Part III will examine how to help conservation go viral—how to spread the ideas of the carbon code so that once you've minimized your own footprint, you can help others to do the same.

SUMMARY

➼ The **carbon code of conduct** helps you make smart decisions about your use of carbon. It is designed to help you make and stick to a personal climate commitment—a commitment to making a positive contribution to the global fight against climate change.

➼ The carbon code has four Rs:

➼ **Reduce**—Do less of the things that emit carbon. *Take fewer flights. Drive fewer kilometers. Eat less beef.*

➼ **Replace**—Do the same activities, but switch to versions that use less carbon. *Take the train instead of a plane. Drive efficient vehicles. Buy electricity from clean providers.*

➡ **Refine**—Get as much benefit as possible from each unit of fossil fuels that you burn. *Combine a vacation with a business trip. Stay at places that you travel to longer, and do more when you're there.*

➡ If you've satisfied all three Rs, then **rehabilitate** the atmosphere through the purchase of gold standard carbon credits (see chapter 7).

➡ If you are burning fossil fuels in the name of conservation, then your activity should pass an additional test: On balance, will your use of carbon be likely to produce a net reduction in humanity's carbon footprint or a net increase in the planetary resilience to climate change? Be prepared to articulate your reasons publicly.

➡ Some activities may fit in multiple Rs. That is okay. The carbon code is a decision-support tool, not a rigid rulebook.

➡ Avoid information paralysis—any action to reduce GHGs is better than no action. Do not let incomplete understanding of the issue prevent you from taking steps to reduce your personal footprint. Engage in conspicuous conservation. Adopt the carbon code of conduct in your own life. Design a carbon code for the institutions in which you work. Support businesses that operate with a carbon code of conduct. Vote for candidates who support carbon-friendly policies and who apply the carbon code to their own campaigns.

LIVING BY THE CODE

ossil fuels are incredible. They are versatile, packed with energy, and cheap. But like so many good things, they are best in small doses. To build a responsible, low–greenhouse gas lifestyle, we have to phase fossil fuels out of our lives. Getting to that point will be a long, hard journey. Fossil fuels permeate every aspect of society. So how should we prioritize the reduction of our personal climate footprints? How can we best implement the carbon code?

In Part II of this book, I explore the activities that are responsible for the biggest amounts of our GHG emissions. I focus on the things that we do every day. Setting good habits, which are repeated day in and day out, is critical to reducing your carbon footprint. The carbon code is not about restricting you—it is about empowering you to live a clean, sustainable life, while doing your part to solve climate change. And the changes we outline in this part will help you do just that.

We will avoid information paralysis. We will not agonize over details—any one of the topics we're about to cover could warrant a book of its own. Rather, part II is a broad introduction to sustainable behaviors and technologies that we should each consider for adoption. Not every option will be appropriate for every person, but adopting even one of these recommendations—and telling others about it— will make a difference. If everyone who reads this book does that, our collective impact will be substantial.

In this book, we're also going to avoid science fiction. Any technology that we discuss here will be something that either exists right now or that is scheduled to come to market in the very near future. Technologies that may have scientific merit but are many years away from commercialization will be mostly ignored. To fix climate change we have to act now, while simultaneously looking for better solutions as they present themselves.

According to the United Nations' International Panel on Climate Change, 25% of humanity's GHG emissions were associated with electricity generation in 2010, including heating. Consequently, electricity will be the basis for chapter 4. Chapter 5 will look at GHGs resulting from transportation, which make up 14%. Agriculture and land use make up another 24%, and these are tied directly to the choices you make at the supermarket. Hence, diet will be the subject of chapter 6. Finally, chapter 7 will examine long-distance travel—a subset of transportation's 14% but one that those of us from rich countries are disproportionately responsible for.

Collectively, these activities make up 63% of the world's emissions. The remaining 37% of GHGs are shared among buildings, industry, and other GHGs from the energy sector—we can influence these but not directly control them with our own actions. Therefore, part II will focus on elements of that 63% that we can directly do something about. By the time you're done reading this section, you should have some ideas about how to reduce, replace, and refine your carbon usage and will understand how best to rehabilitate as well.

ELECTRICITY

The fundamental problem behind climate change is that we rely on technologies that are tied to the emission of GHGs. As I said in the prologue, a full quarter of our GHG emissions come from our use of fossil fuels to make electricity. In the United States, it's even more than that: a full third goes to electrical generation.

Every time we burn fossil fuels, chemistry dictates that CO_2 is released as a waste product. No matter what labels we attach to the fuel—for example, by slapping the word "clean" in front of coal or diesel—GHGs are produced. While some fuels are worse than others, every unit of energy that is derived from fossil fuels produces carbon pollution. It follows that transitioning to a carbon-neutral world requires that we must mostly eliminate the use of fossil fuels to power our lives. With a quarter of the world's carbon emissions tied directly to electricity production, any plan to solve climate change must address our production and consumption of electrical energy.

Let's be clear—without electricity, there simply would not be seven billion of us. Electrical energy keeps our food and water systems working, our industries active, and our global finance network humming away. It's not a convenience. Electricity is a core service that we depend on. It's also something that many people in the world deserve to consume more of to increase their standards of living to what is taken for granted in wealthy countries. If we value social justice (which we should) then we also have to accept that empowering people to lift themselves out of poverty requires the availability of affordable electricity. We built our civilization around cheap, abundant, dirty energy. Helping developing nations get to the same place we are without following the same energy pathway should be an absolute priority.

Unless you're in charge of your city's energy utility, you probably don't get to choose the way your power is produced (though you can certainly influence it—we'll talk about that in part III). However, there is still plenty you can do to reduce the amount of electricity you consume and thereby reduce your personal carbon footprint as well.

In chapter 3, you made a big decision to take action in support of the climate. In this chapter, we're going to discuss how that big decision can affect the way you use electricity. Reducing and refining electricity usage is your first step to reducing your footprint. Replacing dirty energy with clean energy is the second step and involves purchasing clean power, as well as generating and storing energy at home. We'll cover all this and then conclude the chapter by discussing the pros and cons of the various types of electricity production that currently exist today.

A PRIMER ON POWER AND ENERGY

Before we discuss the carbon footprint of electricity, it's important to understand some basic terminology. **Electricity** refers to the movement of electrons, or tiny charged particles that are associated with atoms. **Energy** means the capacity to do work and exert a force over a distance. Therefore, **electrical energy** is work done by the movement of electrons. Energy can be measured in many different ways, including British thermal units (BTUs), joules, and calories. When we talk about electrical energy, we use the standard term of kilowatt-hours, or kWh. One kWh means one kilowatt of power maintained for one hour. One kWh is equal to 1,000 watt-hours, or Wh.

Energy is a quantity that you can add up, like a liter of water or a gallon of milk. When you get your electricity bill, you are usually charged by the kWh. The electrical utility counts how many kWh you used over the billing cycle and then charges you accordingly. In the United States, most people are charged between 8 and 17 cents per kWh for their electricity, and the average home consumes about 900 kWh per month.

Power is different from energy—power is the **rate** at which energy is produced or consumed. Every electricity-powered device in our homes requires a certain amount of power to operate. The unit of power is the watt (W), and 1,000 watts is a kilowatt (kW). Other common units you may hear are megawatts (MW—a million Watts), gigawatts (GW—a billion W), and terawatts (TW, a trillion W).

Let's put all these terms together into a real-life example. If you install a 100 W lightbulb and leave it on for 1 hour, you will have consumed 100 Wh of electrical energy, or 0.1 kWh. If you leave it on for ten hours, you will have consumed 1,000 Wh, or 1 kWh of energy. Your bulb consumes 100 W of power, and your power company will bill you for 1 kWh. If a kWh is like a liter of water in a bucket, then think of kW as meaning that the bucket fills at a rate of one liter per hour.

Another important principle is that power can be produced or consumed. You can have a 100 W lightbulb, which consumes 100 Wh of energy if left on for one hour. You can also have a solar panel that produces 100 W of power—this panel would generate 0.1 kWh of energy in an hour, at full capacity. Energy is a quantity, whereas power is a rate.

Now that you understand the difference between power and energy, you can start to look at how to reduce the overall amount of electrical energy that you consume—and thereby reduce the amount of carbon you're responsible for emitting through your electricity use. The goal here is to replace high-power devices with devices that require less power to operate, to reduce the amount of time those devices are used, and to refine the way we use these items to get more benefit for the electricity we consume. Ultimately, your electricity-related carbon footprint is determined by the number of kWh you consume and the way those kWh were produced. We'll talk about consumption first.

THREE RS AND ENERGY CONSERVATION

Reduction, replacement, and refinement govern the three prongs of energy conservation, and all three work together to reduce the number of kWh that we use every month. We can reduce the number of items that we use or use them for shorter periods of time (e.g., turn off our aforementioned 100 W bulb when we're not using it). We can replace inefficient items with devices that do the same thing, but at a lower power (e.g., replace our incandescent bulb with something more efficient). In addition, we can refine the way we use items so that we get more bang for our buck (e.g., by purchasing a luminaire that spreads light better in our room, or by painting the room's ceiling white). Here, refinement implies that we'd be able to reduce our overall energy consumption due to its smarter use.

Let's unpack these ideas a little further. Energy conservation is partially behavioral. This means that you need to employ simple tactics to keep the amount of electricity you consume to a minimum. Every dollar you do not spend on electricity is a dollar you can spend on something else, like savings, rent, or a mortgage. So you have every reason to do this. And if you've made the big decision to follow a carbon code of conduct, then the answer to the little question of "Should I conserve energy where I can?" becomes a clear yes.

Imagine that each electrical device you own is a hose attached to a tap. When you use the device, it's like turning the tap on full. Energy-intensive devices are like big hoses, while efficient technologies are like smaller hoses. Regardless of the size of the hose, they make a mess when you leave them on all day, so your first goal should be to make sure that you only use devices when you actually need them.

Let's start with the easiest behaviors. Lights make up about 14% of the energy consumption of a typical home in the United States. This may seem obvious, but lights only need to be on when you need to see. That means when you leave the house, make sure you remember to turn those lights off! When you're in the house, only leave lights on in rooms that you're inhabiting. To do this, you have to develop a habit of turning off lights when you leave the room. When you're leaving your house for the day, switch off the outside lights. It's an easy habit to get into, and it can save you lots of money and carbon over the long term.

The next thing to consider is your heating or cooling. If you live in a cold climate, then heating is probably the biggest source of power consumption for your home. If you live in a hot climate, then air conditioning may play a major role in your energy bill. Reducing your expenses in either category means keeping the thermostat a little closer to the outside temperature and ensuring that your home is well insulated.

The simplest way to reduce consumption of energy for heating is to keep your house at a reasonable temperature, no warmer than 21°C (69.8°F). According to the American Council for an Energy-Efficient Economy, you can save about 6% on your heating bill for every degree C cooler that you keep your house. This figure is simplistic, since it depends on how warm your house was in the first place, as well as what technology you use to keep your house warm—but the principle is absolutely correct. A lower internal temperature means less energy used, and that in turn means a smaller carbon footprint. This is particularly important if you like your house very warm. It takes disproportionately more energy to heat houses to higher and higher temperatures. The reason is simple physics. Temperatures don't like to be different. When two objects come into contact, a larger temperature difference will result in a faster rate of temperature change between two objects that come into contact. Imagine throwing an ice cube into boiling water—the cube quickly disappears. By contrast, put that ice cube in cold water and it will take a while for the ice to melt.

The same is true of the air in your house. A very warm house in a very cold climate will want to lose heat to the outside world at a higher rate. My grandparents have this all figured out. They live in Vancouver, and while it's a relatively warm city by Canada's standards, it is certainly not balmy in the winter time. It's rainy, damp, and moist for at least half the year—yet they do not turn their heater on unless it is absolutely necessary. I can remember Sunday dinners where we would have to bring sweaters, mittens, and hats to make it through dessert without losing the feeling in our hands. Their commitment to conservation was admirable (although I must point out that it is not recommended for house temperatures to drop below 16°C (60.8°F), as it can promote moisture build up and ultimately mold growth).

Another contributor to the household energy footprint is the suite of large appliances that are found in most modern homes. These include the dishwasher, the clothes washer and dryer, the refrigerator, and more. When selecting these appliances, it is best to look for ones that have been endorsed by Energy Star. This is a program run by the United States Environmental Protection Agency, which identifies products that are energy efficient. Beyond selecting certified appliances, there are other small things you can do to save energy as well. For example, a full fridge consumes less power than an empty one. The reason is that food (which contains a lot of water) is more resistant to changes in temperature than is air. This means that when you open a full fridge, the compressor doesn't have to work as hard to re-chill the space. Likewise, for washing machines, driers, and dishwashers, you should make every effort to run them as full as possible. They work more efficiently, and you will save money on energy, detergent, and wear and tear on the unit.

Other electronic devices typically make up another substantial chunk of a household's energy use. These include televisions, computers, and the vast array of chargers that power our cell phones, tablets, and other gadgets and gizmos. While most modern electronics are efficient, many of them also bear a "phantom load," meaning they consume energy even when turned off. Set-top digital cable boxes are among the biggest culprits here. By some estimates, cable boxes are the single biggest consumer of energy in many households, behind heating or air conditioning. Many boxes consume about 50 W whether they are off or on, and they do so 24 hours a day, 7 days a week. A 2014 article in the *Los Angeles Times* estimated that it would take four nuclear reactors to power all the cable boxes in the United States—just for this one single appliance in so many modern homes.

With set-top boxes you're basically stuck, because unplugging them means a lengthy reboot period every time you try to turn them back on. My wife and I got rid of ours altogether, partially for this reason. Your best strategy with electronics is to get in the habit of turning them to the lowest activity setting possible. Turn them off when you're not using them, rather than leaving them on standby. If it's a device like a cellphone charger, then unplug it when it's not in use. Keep them out of the wall, and you'll save a few dollars on your power bill over the long term.

There are tools available to help you fine-tune your electricity consumption. The Kill-A-Watt meter is a neat device that gives you a digital reading of the consumption of any device that can be plugged into a wall outlet. This is great for measuring phantom load and seeing which devices in your house consume the most energy. Assessing your whole-home power consumption can be done directly, by just looking at your bills over time but also by purchasing a home energy monitoring kit. These are devices that give you real-time readouts of your electricity consumption, so that you can modify your behavior to reduce it.

SAVING MONEY BY SAVING ENERGY

Replacing inefficient technology with energy-efficient devices is the next step to energy conservation. If you're like me, and you enjoy interesting new technology, this is the fun part of conservation. Any time you buy something new for efficiency reasons, it means you're spending money to save energy. In the long run, these technologies can pay for themselves by saving you money on your electricity bill. How long it takes before this happens is known as the payback period.

The payback period refers to the amount of time it takes for a new technology to pay for itself after you buy it. In terms of energy conservation, payback period is determined by four factors: the difference in purchase price of the efficient versus the inefficient device, the difference in power consumption between the inefficient and efficient units, the cost of electricity, and the amount you use the item each day.

For example, one of the best investments for power conservation is to replace incandescent light bulbs with light emitting diodes, or LEDs. These are lights that look just like normal bulbs but consume a small fraction of the energy to get the same brightness and a similar quality and color of light. However, unlike a normal 60 W bulb, which costs about $0.50 to buy, a replacement 9.5 W LED might cost $8. Is it worth making the switch? Let's do the math.

At my home in St. John's, in the province of Newfoundland and Labrador, Canada, electricity costs about $0.11 per kWh. My current bulb is 60 W (0.06 kW), meaning it costs about $0.007 per hour to operate ($0.11 per kWh, times 0.06 kWh). An equivalent LED, at 9.5 W, costs $0.001 per hour to operate—so we're saving $0.006 per operating hour with the LED bulb.

The LED costs $8 to buy, and the incandescent $0.50—so there's an additional cost of $7.50 for the efficient technology. So how many hours of light usage does it take before we save enough money to pay for this additional cost (recall that we save $0.006 per hour of operation)? The answer is $7.50 divided by $0.006 per hour, or 1,250 hours. In other words, once we've used this new LED for 1,250 hours, it has paid for itself in savings, and anything beyond that is money in our pocket.

But how long will it take us to reach 1,250 hours of use? This depends on where in the house we use the light. In my living room, I use my lights a lot, about 8 hours per day in the winter. Therefore, my LED would pay for itself in 1,250 hours divided by 8 hours per day, or 156 days. In less than half a year my light will have paid for itself, and I'll be saving money thereafter! This estimate is conservative because LEDs are rated for about 50,000 hours of use, compared to 1,200 hours for incandescent bulbs. In other words, over the lifespan of that LED bulb, I'd have had to buy a new incandescent bulb 41 times.

Now that I've walked you through the math, you can see how the four factors affect the payback period in this example. Let's say I lived in Hawaii, and power cost me about $0.33 per kWh. In this case, my LED would be saving me almost 2 cents per hour versus the incandescent bulb—at 8 hours a day of use, it would pay for itself in only 47 days! Sticking with Hawaii, let's assume LEDs cost only $4 rather than $8—in that case payback occurs in about 22 days.

This example demonstrates the fundamental challenge of switching to efficient technology. The financial argument is most compelling when energy is expensive, when the device is used a lot, and when the difference in cost between the efficient and inefficient technology is minimal. This shows why it's important to make sure that dirty energy and inefficient technology do not get any subsidies from government. Promoting the shift to better technology means that we need to, at minimum, allow the free market to dictate which technology is favorable. Better still, we should recognize the fact that clean technology is in the public interest and shift subsidies onto these devices so that people adopt them faster (we'll talk about this in part III).

However, no matter how cheap energy-efficient technology is, no one will use it if it's not a good alternative. In the case of LEDs, they are excellent replacements for traditional bulbs. Some LEDs emit a "warm" color that is lower in color temperature, similar to regular bulbs—or, you can go with an intense white light if that's what you prefer. LEDs are also dimmable, which is important if your light has a dimmer switch. Another option in lighting is to use compact fluorescent lights, or CFLs. CFLs were a transition technology between incandescents and LEDs. They're cheaper initially than LEDs and last for about 8,000 hours as opposed to 1,200 for incandescents. They consume about twice as much power as an equivalently bright LED—still far better than incandescents, but not as good as LEDs. However, many people feel the quality of light emitted by CFLs is not as good. In addition, CFL's don't work on dimmers, unless you spend extra money to get a dimmable CFL (at which point there is literally no reason not to just get an LED). All this being said, LEDs are better in so many ways that I wouldn't recommend bothering with CFLs. Just make the jump right to LEDs and you'll be more satisfied overall.

There are a few myths that persist about lighting, which we can address directly here. The first myth is that they produce an ugly color of light. As mentioned earlier, make sure to buy the "warm"-colored lights and they will be nearly indistinguishable from incandescents, especially if they are behind a shade or in a light fixture. Second, there is a myth that these lights take a long time to warm up before they are fully bright. That's not true at all for LEDs, which achieve full brightness immediately on activation. Some early CFLs had this problem, but modern ones do not. Third, there's concern that all energy-efficient lights contain mercury. That's not true. LED bulbs do not contain mercury. However, CFLs do—so be very careful if a CFL breaks in your house, and follow appropriate disposal directions.

But let's get back to the #1 expense on most people's energy bills—heating and cooling. To be clear, heating and cooling are not always electric. You may burn natural gas, propane, or even fuel oil to heat your house. But for the purposes of understanding conservation, that distinction doesn't matter all that much. The less you heat or cool your house, the less energy you use, and the lower your carbon footprint is.

The first goal should be to minimize air exchange between the inside and outside of your home. That means making sure the building is well insulated. Start by looking for gaps in your home that would allow for draftiness. For example, examine your doors and windows to see if there are any visible gaps. Where you can see daylight, you will want to purchase weather stripping and install it. This costs only about $15 per door and will greatly improve comfort in the home. If you live in a cold climate, covering bare floors with area rugs is a great way to make the place seem warmer. This will reduce the amount of heat you lose through your feet onto the floor. Get window coverings that keep the heat in as well, and install double-paned windows where you can. Try to keep your use of space heaters to a minimum, as they are very energy-hungry.

If you're building a new home or replacing an existing furnace, consider installing an outdoor heat pump instead of a traditional furnace. Heat pumps work by sucking up heat energy from the outside world (even when it's cold out!) and pulling it into your home. Rather than generating heat, they move it from outside to inside. They can also work in reverse, gathering heat from inside your home and moving it out. Another great option from a climate perspective is wood pellet furnaces, if they are available in your area. The carbon footprint of these is very small, and the pellets are cheap to purchase.

If you're lucky enough to live in a warm climate, minimize use of air conditioning unless you have a medical reason not to. Most of the world lives without A/C, and it consumes a staggering amount of energy. Central air conditioners consume about 3.5 kW, while window units consume around 1 kW. By contrast, a ceiling fan on high consumes about 75 Watts, or 0.075 kW. It's important to see A/C for what it is—a convenience, rather than a necessity, for the majority of households. The reality is that in most warm countries, air conditioning in homes is relatively rare. In the United States, it is surprisingly common, and this represents a major source of energy consumption. China is starting to see more adoption of A/C, and this is scary, given that country's large-scale use of coal power plants.

Neither heating nor air conditioning needs to be on full blast all the time. Your options here are to turn down the thermostat when you don't need heat (at night, or before you leave for the day) and then turn it back on when you need it. If you're using a heat pump, then you're better off leaving the thermostat at a stable level, as it consumes extra energy to heat a space rapidly compared to keeping temperatures constant. But all HVAC technologies can be improved upon by using specialized thermostats to control your heating and cooling. Programmable thermostats represent one option. These are devices into which you can program a schedule for your home. For example, if you typically wake up at 7 a.m., you can set it to start warming the house at 6 a.m. In doing so, you're only maintaining the higher heat level when you need to, and you save energy at other times of the day.

A second type of thermostat, which is a step up in terms of features and capabilities, is a smart thermostat. In addition to being programmable, these devices have motion detectors that enable the onboard computer to "learn" your behavior, so that the device can fine-tune your home's energy consumption to maintain it at the right level. When you're away, the house will cool, and when you are home, it will warm up. Smart thermostats are very easy to install—an online guide shows you how to do it step by step, and you can do it yourself if you're even slightly handy.

Speaking from personal experience, when my wife and I replaced our programmable thermostat (which wasn't really suitable for use with our heat pump) with a smart thermostat, our heating bill dropped by about 25% year over year, despite the fact that our winter was actually colder that year. You can also control these thermostats from an online app, so if you know you're going to be coming home early, you can use your smartphone to turn up the heat. Smart thermostats are usually between $200 and $250. Check with your utility company to see whether they have rebates. In our case the payback period was about a year, given how much we saved on our energy bills.

GENERATING ELECTRICITY AT HOME

To solve climate change, it's important to make sure we don't waste our kWh. However, efficiency on its own won't completely solve the problem. In fact, as we phase out things such as gas-powered cars and go electric (more on that in chapter 5), we may actually increase our electricity consumption. Therefore, it's just as important to make sure that the kWh we consume are produced in carbon-neutral ways.

Thanks to rapid decreases in renewable energy costs, it is now possible to install systems to generate electricity right at your own home. Let's start by discussing solar panels. When you picture a solar panel, you're probably thinking of a photovoltaic (PV) panel, which is made up of many small cells that convert light energy into electricity. These panels are scalable, meaning that many small panels can be put together to equal the production of one large one, and there is no limit to the number or size of panels that are deployed. The design of PV panels is constantly improving, and their price is rapidly declining.

The other common type of solar power generation works through concentrated solar power (CSP) systems. These systems employ arrays of mirrors that focus light energy onto a single point. Often that point will be on a tower in the middle of a large circle of mirrors. This incredibly intense beam heats up water that subsequently powers a steam turbine, generating electricity. Unlike PVs, CSP is only built at large scales.

You can get a lot of electricity from solar panels. Consider this: British Columbia, Canada, has been trying to build a new major hydroelectric dam called Site C for many years. The most recent estimate suggested that it would produce 1,100 MW of power, at a cost of $9 billion. This dam would flood an entire river basin, destroying agricultural land and displacing aboriginal people. Millions and millions of dollars have been spent getting the project to its current state, and aspects of it remain tied up in the courts. For the foreseeable future, it will generate no energy, yet it still costs taxpayers money.

By contrast, Morocco is building its own $9 billion mega-project. It is a solar power plant and will be a mixture of PV panels and CSP. Its energy generation will put Site C to shame. Two thousand MW will be generated with a relatively minor environmental impact, compared to a major hydroelectric project. While birds and other flying creatures will be the primary victims of CSP plants (they can get burned if they are hit by concentrated solar beams from mirrors), the cumulative impact of this is low compared to nearly any other energy generation option. In the case of birds, the number killed annually by housecats, vehicle strikes, or even office tower windows dwarfs that of even the most impactful CSP plants. Even wind turbines, which are often opposed based on their perceived negative effect on birds, may not be overly damaging in comparison to other energy-generating technologies. In 2013, an article in the science journal *Renewable Energy* examined the number of birds killed per GWh of energy generation and compared results by the type of power plant. The paper showed that for every GWh produced by wind farms, 0.3 birds are killed. By contrast, fossil fuel plants killed an eye-watering 5.2 birds per GWh, totaling more than 14 million dead birds in a given year. You wouldn't want to put CSP plants, wind turbines, or any other technology in critical bird habitat, but the data so far suggest that well-sited renewable plants can be a win-win for biodiversity and climate change, especially in comparison to the nonrenewable technologies we currently depend on.

But this book is primarily about what you can do as an individual. And the beauty is that if you own your own home, you can probably install some amount of solar energy equipment. With solar energy, you can have a tiny panel that powers your calculator, a massive megaproject that powers a country, or anything in between. Solar panels can generate electricity even in cloudy conditions, though they obviously generate more in sunnier climates. In an effort to promote home solar systems, Google has introduced Project Sunroof,[1] an online application that calculates the amount of solar energy you could expect to generate at your home.

Installing your own solar PV system isn't quite as simple as buying a panel and plugging it in. Solar panels have to be professionally installed and connected to your household's energy systems. Generally, this can work in three ways. The first and most common is to enter into a "net metering" arrangement with your power company. Here, your PV panel connects to the electrical grid, and any energy that you generate is sold back to the power company. Proceeds from this sale are then deducted from your electrical bill. Net metering is currently the most common approach to home generation. In the state of Hawaii, so much energy has been generated this way that the utility has desperately been trying to reduce the amount it has to pay for this type of electricity. Nearly one in eight homes in that state has home solar generation. If your jurisdiction doesn't already allow this, you should pressure them to do so.

The second method is to not use a PV system at all—rather, you could install a solar thermal system that uses the energy of sunlight to directly heat water in tubes encased in a large panel. The idea here is rather than having complicated electronics that get the electricity ready for use in your home, all the energy goes directly to heating. This generally wouldn't be the sole heater for a home, but it could reduce the amount of energy that you consume on a sunny winter's day.

The third way to take advantage of home solar is to go completely off the grid, by adopting both PV panels and energy storage. In this model, your house would have large batteries connected to your PV panels. Your panels would charge the batteries, and the batteries would power your house. Three components are required for this to work: the PV panels, the batteries, and an inverter that converts direct current (DC) battery power to the alternating current (AC) power used by electronic devices in the home.

There are a number of products coming to the market to support home energy storage. The electric car company Tesla Motors (we'll talk a lot more about them in chapter 5) has gotten into the energy storage business with a product called the Powerwall, a home battery designed to be repeatedly charged and discharged throughout the day. The Powerwall has enough power to operate a typical home for several hours. If you need more energy, you can string up to nine of these devices together and remain off the grid for quite some time. During the day, while you're out of the house and not consuming much energy, the batteries recharge using solar panels.

Energy storage can save you money on your power bill in other ways, too. Many jurisdictions charge different rates for power consumption at peak times of the day. This is called "time-of-day pricing" and is in place in much of Europe. For example, in the evenings, when everyone is home cooking, doing laundry, and so on, the price per kWh for electricity might be higher than at other times of the day. If this applied to you, then your home batteries could charge themselves at night, when power is cheaper, and discharge during peak hours. In addition, there are obvious benefits during power failures—your lights would stay on while others' go dark.

At present, home generation and storage isn't cheap. But as it becomes more mainstream, its price is certain to drop, just like we have seen with other renewable energy technology. It wouldn't surprise me if, 10 to 20 years from now, home battery systems are as common as water heaters, laundry machines, and other appliances that we take for granted in a modern home today.

For home energy generation, there are two other options that can be workable in specific circumstances. The first is wind energy, where we convert the kinetic energy of moving air into electrical energy. As with solar, home wind generation is only workable by either charging a battery or through selling the energy back to the grid through grid-connected systems. Unlike with solar, wind energy is only economical for home use in very specific situations. You have to live in a place that is very windy. According to the Wind Energy Foundation, you need at least an average wind speed of 16 kph (10 mph), or about 4.5 meters per second, to make a wind turbine economical. This is faster than you'd think. To determine whether you live in an area with this average wind speed, you can check your country's wind atlas.

Wind turbines work best when they are really tall and when their blades are very wide. The size requirements make them less economical for home usage than solar equipment. Installers typically recommend that home turbines be at least 25 m (82 ft) in height—about twice the height of a telephone pole. The diameter of the blades should be as large as you can get, to suck as much energy from the wind as possible. Depending on where you live, this combination of requirements may or may not be reasonable. But, if you live in a windy place and you have a lot of land, it's certainly something to consider, and a wide variety of products that range in price and generating capacity exist.

A third option for home generation is geothermal energy. At the utility scale, this energy source can be used to generate electricity, but for a private home it is used only for heating or cooling. Geothermal systems are composed of water-filled tubes that are placed deep in the ground. While air at the surface varies in temperature throughout the day or at different times of year, the temperature below the surface is relatively constant. Tubes put into the ground experience this consistent temperature. When it's cold out, the heat from the ground leeches into these pipes, whose water is then pumped to the surface and run across a blower.

In other words, in geothermal systems, you don't generate heat, you move it from one place to another. You also don't directly generate electricity. Rather, you reduce the amount of energy you need to heat your home, which is typically the biggest source of energy consumption in a residence. Geothermal energy is extremely efficient—for every unit of energy spent, you get 4 units of heat energy in return. By contrast, a typical forced-air furnace gets about 0.92 units of heat energy per unit of energy consumed.

Geothermal energy systems are not cheap to install, and they require professional installation. A 2,500-square-foot house will require a system that costs anywhere between $20,000 and $25,000, depending on its configuration. The components inside your home can be expected to last 25 years, and the pipes outside can last as long as 50 years. Depending on energy prices, the payback period could range anywhere from 2 to 10 years. Geothermal systems require a fair bit of land to install the piping.

If these options are all too costly, but you still want to become less dependent on dirty electricity, then there may be one other option available to you. In some jurisdictions, companies exist to sell clean energy to customers. The energy is still delivered through the same power grid, but it is produced in a renewable way. An example of this is a company called Bullfrog Power in Canada. They obtain energy from a variety of clean sources, such as wind. Here, you're paying a premium for clean energy, and what you get in return is the knowledge that your electricity is not coming from fossil fuels. For dealing with your own footprint, replacing more-damaging electricity with less-damaging electricity is a great step to take.

POWERING THE CARBON-NEUTRAL FUTURE

This book is about what you can do as an individual to fix climate change, but it's also important to be energy literate in a broader sense. The first half of this chapter covered what you can do directly, but the rest of it will examine the many ways we generate electricity at the industrial scale. The manner by which we produce electricity impacts our climate and the biosphere, and understanding the tradeoffs of the more common methods will make you a better climate advocate.

I'll start by going straight to the punchline: the best technology pathway to a truly green electrical grid is through the widespread adoption of solar- and wind-based energy systems, complemented by distributed energy storage. This is the only combination of technologies that has become consistently cheaper over time and that can be deployed at sufficient scales, with manageable environmental impacts, to eventually displace the dirty energy sources that we rely on today.

Let's take a step back and remind ourselves of one of the key facts about electricity generation in the context of climate change. Stopping climate change requires that we stop burning fossil fuels for electricity. This will be a long process. Our world is not set up for it right now, and for a time we will still need fossil fuels. However, the fact that we need fossil fuels now is not an excuse to make decisions that further increase our dependency on this resource. We must produce an abundance of energy in a way that does not release CO_2 or other greenhouse gases into the atmosphere. The production should have a minimal negative effect on local ecosystems as well. The cost effectiveness of this shift, over the short term, should be considered but will ultimately be less costly than failing to act on the climate problem. We can invest now, or we can pick up the pieces later.

This concept—that we need to shift our infrastructure entirely away from fossil fuels—is critically important. We build things to last, so if we build more fossil fuel power plants, we're stuck with them for a long while, and every year they operate, they further exacerbate the climate problem. Every time we invest in a piece of infrastructure, whether it's a solar plant, an oil pipeline, or a coal mine, we will have to live with it for an extended period of time. If we suddenly change course and decide to decommission something that has many usable years left, then making a shift will be far costlier. This is why we must be careful and avoid making the mistake of investing in more energy infrastructure that is dependent on fossil fuels.

The vast majority of power plants are called "thermoelectric plants," and they all generate electricity in basically the same way. In these plants, a fuel source is used to produce heat, which warms water until it turns to steam. The steam moves a turbine, which generates electricity. After doing this, the steam is chilled over a condenser, causing it to transition back into water, which can be re-warmed by fuel to make the cycle begin again. The boiler is a closed system, while the condenser is not—cold water has to be constantly fed into it to keep the system cool and turn steam back into water. The water expelled by the plant is not polluted chemically, but it is heated by its contact with the power plant. In other words, to make energy, fuel and cold water go in, electricity and warm water come out—along with the waste products associated with the fuel that power plant uses. Collectively, about 80% of the world's electricity supply is produced in thermoelectric power stations.

Despite the fact that these stations operate on similar principles, their impact on the climate varies widely and is determined mostly by the type of fuel that the power plant consumes.

Let's start with the worst one—coal-fired power plants. Coal is the most abundant fossil fuel on Earth. Burning it for electricity is an absolute catastrophe for the environment. If I could wave a magic wand and immediately remove one thing from our energy mix, it would be coal. It's frankly absurd that we still use this fuel in such a widespread manner. Burning coal for energy made sense during the industrial revolution, when our other option was to chuck logs into a fireplace. Yet shockingly, it still produces about one-third of the United States' domestic energy supply.

First, coal is awful from a CO_2 perspective. The US Energy Information Administration keeps statistics on how many pounds of CO_2 are emitted per million British thermal units (MBTUs) of energy released. Coal emits between 205 and 229 pounds of CO_2 per MBTU. Diesel and gas emit 161 and 157, respectively. Propane emits 139, and natural gas is responsible for 117 pounds of CO_2 per MBTU. So to produce the same amount of heat, coal releases about double the GHGs as natural gas.

The American Lung Association demonstrated in a 2011 report that coal plants produce more hazardous air pollution than any other industrial source in the United States. The particulates that emerge during coal burning degrade air quality and trigger asthma and other respiratory illnesses. Being an asthmatic myself, I am particularly troubled by this. There's nothing quite like having your airways squeeze shut during an asthma attack to make you appreciate a good, clean breath of air. Widespread burning of coal is one of the reasons that many cities in China are suffering from air quality so bad that it has forced public health bodies to issue advisories to stay indoors. As many as 500,000 people die per year in China due to air pollution, and many more are made ill. To clean up the air in advance of the 2008 summer Olympics, officials in Beijing had to close most of the region's factories and power plants for the duration of the games.

But coal plants don't just degrade our air quality, they also deposit all kinds of toxic compounds into the rest of the environment. Mercury is an example of such a toxin. It's a heavy metal that is a major byproduct of coal power plants and is poisonous if ingested. You've probably heard to be careful when eating tuna, because it can be high in mercury. Well, this mercury originally came from emissions from coal plants. These plants also contaminate the environment with low levels of radiation, as the ash is slightly radioactive. When you calculate the sheer volume of ash that comes out of a single plant, it turns out that a coal-fired plant releases 100 times as much radiation into the environment as an equivalent nuclear power generation facility.

You may have heard of something called "clean coal." This marketing concept refers to a variety of techniques that remove some of the more hazardous compounds that come from a coal plant's smoke stack. For instance, they reduce the amount of sulphur dioxide, or SO_2, that escapes into the atmosphere when coal is burned. SO_2 is a critical ingredient in the toxic soup that produces acid rain, so reducing it is certainly positive. But "clean coal" does nothing to lessen the carbon footprint, nor does it reduce the environmental impacts of the mining process that collects the coal for use in the plants in the first place. Carbon capture and storage (CCS) is another option for making coal less damaging. Here, the CO_2 is captured from the plant's emissions and transported to some location where it can't escape into the atmosphere. However, CCS takes energy, meaning that you actually have to burn even more coal to get the same amount of usable electricity, because you are spending energy moving the CO_2 into storage. With additional research and development, CCS could play a part in the solution to climate change. But regardless, the best thing we can do right now to reduce CO_2 emissions associated with electricity is to phase out coal as quickly as possible.

There is one caveat here. Some coal is used in industrial processes, such as melting iron ore to make steel. This represents a use of coal that can help us transition toward a clean energy future. It takes a lot of coal to make a wind turbine, for example, since a lot of the construction carbon footprint of the wind turbine goes into its steel and the concrete used to mount it. But this use of coal is by far in the minority. In the United States, more than 90% of coal goes straight into electricity generation.

A small portion of the world's electricity comes from liquid fossil fuels such as heavy fuel oil. Clearly, this isn't a sustainable solution either. It's just as dependent on carbon-producing, nonrenewable fuels and therefore won't solve the climate problem. Surely we can do better.

Natural gas is another fossil fuel that we often use to make electricity. Natural gas is sometimes referred to as a "transition fuel." From a carbon perspective, it's definitely an improvement. Natural gas burns cleanly, with few pollutants aside from GHGs, and it releases less CO_2 per unit of energy than other types of fossil fuel. But here's the problem: a huge proportion of natural gas is derived from fracking.

Fracking (properly known as hydraulic fracturing) is the process used to extract natural gas from deposits located underground. Fracking operations consist of mobile rigs that drill deep down and inject water and chemicals into the Earth's crust at extremely high pressure. This process pushes open underground cracks and fissures, releasing pockets of natural gas. The gas is then captured and able to be used for energy. Proponents of fracking point out that it has made available tremendous amounts of fuel and has reduced our dependency on foreign energy. Opponents cite evidence that the activity can pollute groundwater and even trigger earthquakes. It also uses an incredible amount of fresh water and can consume millions of liters for a single well. From a climate change perspective, it also encourages us to stay reliant on fossil fuel infrastructure—something we have to rapidly transition away from if we're to successfully fight climate change. The full extent of the fracking debate is beyond the scope of this book. Suffice it to say that if someone were fracking near my home or my drinking water, I'd be very concerned. And it absolutely should not be done in any sensitive habitat or in conservation areas of any kind.

While natural gas is more carbon-friendly than other fossil fuels, natural gas harvested through fracking has a slightly larger carbon footprint than traditional natural gas. That's because the extraction process of fracking is far more energy-intensive than conventional natural gas sources. The Tyndall Centre for Climate Change Research in the United Kingdom estimated that it takes about 15 kg of CO_2 per foot drilled for fracking wells—and some wells can be thousands of meters deep. However, as a proportion of all the GHGs that will eventually be emitted by the gas, they estimated that the fracking-related emissions only amounted to as much as 2.9% of the carbon footprint. So while fracking makes the natural gas footprint worse, its primary problems are associated with its other environmental effects, aside from GHG emissions.

So as a fuel source, natural gas is not perfect. I would go a step further and reject the whole concept of a "transition fuel." Why lock more money and infrastructure into the fossil fuel system, when we can invest right now in systems that meet our electricity needs without burning fossil fuels at all?

One way to do this is with hydroelectric power, or large dams. Dams are often held up as an emissions-free method of producing power. But while they don't burn fuel to produce power, they are certainly not without harmful impact.

First, dams actually do release a lot of greenhouse gases, but not by burning fuel. Hydroelectric plants basically work like this: you build a big blockage across a river, and as a result, an artificial lake forms on one side of the dam. This massive reservoir is then fed through turbines, which turn the potential energy of stored water into electrical energy that we can use. The artificial lakes created to power dams tend to cover up a lot of vegetation, which dies and decomposes under water. In the process, it emits lots and lots of methane. As you'll recall from chapter 1, methane is an incredibly potent greenhouse gas— more than 70 times as potent as CO_2.

In 2011, Brazil's National Institute for Space Research estimated that 23% of all the methane released on Earth was due to hydroelectric dams—amounting to 4% of the world's GHG emissions. Their construction is carbon-intensive as well. Most large dams require thousands of tons of cement—and producing 1 ton of cement releases about a ton of CO_2 into the atmosphere (in fact, cement production is one of the most carbon-intensive industries on Earth). Trees often have to be cut down to carve paths for transmission lines, and roads through wilderness fragment habitat and cause further environmental degradation.

It's also worth pointing out that the large headwaters produced by dams can be incredibly disruptive to humans and nonhumans alike. They render uninhabitable large tracts of land, which tend to be the most productive riverfront land, important for food production. They restructure entire ecosystems and cut off salmon populations that historically may have travelled the length of rivers to spawn. In China, the construction of the Three Gorges Dam, the largest single power plant in the world, displaced more than a million residents. Imagine—a city with the population of Dallas having to simply uproot and move to support the building of a power station.

To make matters worse, large dams are ridiculously, prohibitively expensive. A 2014 study by researchers at Oxford University demonstrated that construction costs for large dams typically end up being about double their initial estimates. Furthermore, on average dams take 44% longer to build than forecasted as well. This isn't specific to any one country—the team looked at 245 projects in 65 countries, built between 1934 and 2007. These forecasts were most inaccurate in the poorest countries, which could ill afford the cost overruns. This scenario is playing out in the province of Newfoundland and Labrador, Canada, where the Muskrat Falls hydroelectric dam project has climbed in cost from its initial estimate of $7 billion up to the most recent forecast of $11 billion when all costs are included, and where the year in which the project starts generating electricity has been pushed back repeatedly as well. Dams can bankrupt governments, and yet they are still often perceived as a conservative investment by public utilities.

Finally, even if we decided these costs were acceptable trade-offs, there just aren't enough rivers on the planet to provide for our global energy needs. According to the International Rivers NGO, about 60% of the world's major rivers have been dammed. From both environmental and social justice perspectives, it is probably counterproductive to make it a policy objective to dam the remaining 40%. Even if we did, it would not be sufficient to cover our global energy needs and displace fossil fuels. In addition, as precipitation gets disrupted by climate change, the long-term viability of the headwaters becomes questionable, too.

The other problem with dams is what to do when they become obsolete. Dams don't last forever, and decommissioning them is tremendously expensive. The state of Washington recently decommissioned and removed a large dam on the Elwah River, at a total estimated cost of $200 million. That is a staggering amount of money, especially when you consider how much renewable energy could have been installed just for the price of decommissioning that dam. According to the British Columbia Sustainable Energy Association, it costs about $3.50 per Watt to install solar panels for home use, or between $1.50 to $2.50 per Watt for a utility. Let's be conservative, and use the figure of $3.00 per Watt, or $3,000 per kW for a solar PV installation. At this price, for just the cost of decommissioning this dam, we could have built over 66 MW worth of solar panels (only counting the purchase cost of the panels).

And of course, decommissioning is preferable to the alternative, which is simply that the dam fails and kills everyone that lives downstream of it—a scenario that has happened far too commonly around the world, especially in China. The most frightening example is the collapse of the Banqiao Reservoir Dam in 1975. More than 170,000 people were killed, and 11 million were displaced due to the catastrophic failure of this power plant.

What about nuclear power? This topic is worth spending some time on. It can get pretty heated, with credible, smart people being both for and against this source of energy.

Before we get into this I will disclose my personal view—I don't think nuclear power can play a major long-term role in solving climate change, and I think environmental scientists who advocate for this technology are underestimating its limitations or the challenges in implementing it. That said, in most of this book, I mostly make arguments that represent my interpretation of the scientific consensus. In other words, the carbon code is about focusing on ideas that are mostly not controversial among people who know what they're talking about (but may be very controversial in the mainstream). However, in this case, smart people acting in good faith really do disagree on this topic, so I can't give a simple answer. Nuclear energy is too central to the climate debate to ignore, and there is not a consensus among conservation professionals about whether this technology is a panacea, a disaster in the making, or something in between.

For electricity generation, there are two general categories of nuclear power: fusion and fission. Let's explore fusion first. In fusion energy generation, two hydrogen atoms fuse together to form a helium atom. Due to a trick of physics, the helium atom is slightly lighter than the two hydrogen atoms. In physics, mass and energy are interchangeable, as explained with Einstein's famous equation, $E = mc^2$. If mass is lost when you fuse two atoms, then energy has to be released for the equation to remain balanced. Fusion reactions release incredible amounts of energy—in fact, this type of energy is what fuels our sun.

Currently, there is no such thing as a fusion reactor that generates usable electricity. We don't have the technology to do anything but make bombs or conduct experiments with this type of energy. Fusion reactions have been constructed in laboratories, but the amount of energy it takes to trigger the reaction has far outweighed the amount of energy produced. Might it be possible one day? Sure, maybe. But dealing with climate change, we don't have the luxury of hoping and waiting for an energy breakthrough—we need to use the proven technologies that exist right now, while also researching new renewable energy sources for the future.

The type of nuclear power that we use today is based on fission, where a single atom is split into two, and energy is released in the process. According to the US-based Nuclear Energy Institute, about 11% of the world's electricity supply came from fission plants in 2012. As with fossil fuel power plants, nuclear plants consume fuel to heat up water, which becomes steam and powers a turbine to make electricity. However, the nuclear reaction does not involve combustion. That means no CO_2 or other GHGs are emitted to generate electricity. The nuclear plants themselves do not emit much radiation. In fact, many studies have shown that electricity produced by coal plants causes far more radiation to be emitted into the environment than nuclear plants during normal operation. In addition, nuclear plants themselves have a small footprint on land and can generate a lot of energy on a small site.

Despite the advantages of nuclear plants, many environmentalists have long been opposed to the spread of nuclear power. There is something spooky about nuclear energy—radiation is a mysterious invisible force that mutates cells and gives people cancer. You can't see it or feel it, but you can die if you're exposed to too much of it. And once it's released, there's not much you can really do about it—contaminated landscapes are rendered inhospitable, often for thousands of years.

The permanence of radioactive contamination seems inherently unnatural.

There's also the nasty business that a product of the nuclear reaction that produces electricity is plutonium, which is an important element for making nuclear weapons. Nuclear plants proliferating around the world means more countries would have the means—at least in theory—to make and possess nuclear weapons or sell the fuel to other nations with such ambitions. In fact, if you go back to the 1950s, when acquiring nuclear weapons was a major strategic priority of great powers, the only thing that made nuclear reactors cost effective was to produce electricity while simultaneously selling nuclear weapon fuel to the military.

Practically speaking, people get uncomfortable when despotic or unstable regimes have nuclear plants. I think we'd all feel a lot better if there was less capacity to build nuclear weapons in the world. Proliferation of nuclear plants runs counter to that goal. Plus, it would be awfully silly for us to spend all this time and energy saving the world from climate change if we helped set up the conditions for a nuclear war that wipes out life as we know it.

Full-blown meltdowns are perhaps the scariest side effect of nuclear plants. To be clear, these are extremely rare. To date, the two most serious nuclear accidents occurred at Chernobyl and Fukushima. Chernobyl had out-of-date reactor technology, and that meltdown would not have happened with today's plants. Fukushima's plant was hit by a massive earthquake, followed by a huge tidal wave—in combination, the stresses far exceeded what the building was designed for. However, when meltdowns do occur, it's like winning a lottery from hell. Nuclear accidents tend to kill few people directly, but their indirect effects on cancer rates and other forms of sickness have been the source of much controversy. In the case of Chernobyl, the official death toll was around 4,000, but estimates vary from the thousands to tens and hundreds of thousands, depending on how researchers choose to count the cancer cases that developed in the affected region.

While the actual death rates are disputed, the amount of money spent on dealing with these accidents is not. The after-effects of Chernobyl have been estimated to have cost governments of the world $350 billion, displaced 200,000 people, and rendered a 30 km radius (which contained a town) uninhabitable. To this day, between 5% and 7% of the budget of the government of Ukraine is spent dealing with ongoing issues related to Chernobyl, according to the International Atomic Energy Agency. Fukushima is also an economic catastrophe. The disaster there has cost the Japanese government and the Tepco utility that operated it over $100 billion. Like Chernobyl, it triggered the establishment of an exclusion zone, which remains partially in effect to this day. An additional accident at Three Mile Island in Pennsylvania was much smaller but still cost the US government about $1 billion to respond to. As an aside, I'd remind you that $450 billion could purchase about 150 gigawatts of solar generating capacity at $3,000 per kW—so with the amount of money we've lost due to these accidents, we could have built enough solar panels to generate 34 times the amount of electricity that Fukushima produced.

Another problem with nuclear power is that the energy return on investment, or EROI, is middling at best. EROI is the energy equivalent of the expression "you have to spend money to make money." It refers to the fact that no energy comes freely, and that all electricity generation requires an up-front investment of other forms of energy. For example, petroleum doesn't magically show up in the gas station. Rather, you have to go out, drill for it, transport it to a refinery, process it, transport it again, and then ultimately use it, and each step in that chain of events consumes energy. The EROI of hydroelectric power is about 80. In other words, for every unit of energy you spend building and maintaining a dam, you get about 80 units of energy back. In the case of nuclear power, its EROI has been calculated at about 14—making it competitive with most other fuel sources but not a slam dunk.

The reason for nuclear power's low EROI is that uranium, the fuel of the fission plant, is energy intensive to mine. To make matters worse, as high-quality uranium sources are mined and consumed, we will need to gradually shift toward lower-grade ores that must be heavily processed for use. As a result, the EROI will drop, and the carbon footprint of the fuel source will increase substantially. This was the argument made by Benjamin Sovacool, the director of the Danish Center for Energy Technology, in his 2011 book *Contesting the Future of Nuclear Power*. Unlike renewables, which become cheaper over time, nuclear power appears to be trending in the opposite direction, which is extremely troubling for any potential solution to our climate problem.

The proliferation of nuclear waste is another issue with these plants. Nuclear waste is divided into several categories based on radioactivity level, with the most concerning being "high-level waste," or HLW. HLW mostly comes from the uranium fuel rods that are consumed to produce usable energy. As they get used up, the rods transform into a wide range of metals but remain extremely radioactive. If you were to stand within a few meters of one of these spent fuel rods, you'd get a lethal dose of radiation almost immediately. According to the World Nuclear Association (the industry association that represents the global nuclear industry), about 10,000 m³ (353,147 cubic feet) of HLW is produced each year by nuclear plants across the world, with an additional 200,000 m³ of low- and intermediate-level waste. To be fair, the association also noted that these figures stand in contrast to 400,000 tonnes (440,925 US tons) of ash that is produced by a single coal-fired power plant.

No matter what you think of the volume of nuclear waste, the question remains—it's dangerous and will stay that way for a very long time, so what can we do with it? There are several options that enable the industry to recycle nuclear waste, to use it as fuel in future reactions.

In Europe and elsewhere this is common, but not in the United States. This is a tremendously vexing problem, as nuclear waste that is not recycled has to be dealt with in a specialized way to prevent contamination of groundwater. Our solution was supposed to be a special facility at Yucca Mountain, a deep underground site in the United States that was geologically stable and which could serve as a storage site for nuclear waste basically until the end of time. That facility had its funding cut and was never built. As a result, waste is stored on site at nuclear plants across the country.

Proponents of nuclear power will remind us that when we get rid of our nuclear reactors, we often end up using something worse, such as coal. That is exactly the situation playing out in Japan right now—as nuclear plants are being taken offline, the coal industry is swooping in to fill the gap. By my estimation, that failure is the result of our larger failure to make a decision to deal with climate change in a serious way. If pollution were priced properly to reflect the harm caused by emissions, and if it were not subsidized, the free market would render coal plants financially nonviable. The fact that people make bad decisions in the absence of nuclear power is not evidence in support of nuclear plants. Rather, it's evidence that we need to make better decisions about where we get our energy from.

Proponents of nuclear power also point out that the actual death and disease rate from nuclear accidents is arguably a lot lower than some literature may suggest. Not a single person died in the Fukushima accident, for example. My reaction to that is—that's good, but it doesn't excuse meltdowns as an acceptable cost of doing business. Even if nobody died, the land was rendered uninhabitable, people were displaced, and billions are still being spent on the cleanup. The fact that few people died showed that affected citizens were highly motivated to get the heck out of the contaminated area.

Perhaps the final nail in the nuclear coffin is cost. Unlike most technologies, nuclear power plants are actually getting more expensive to build as time goes on. The Union of Concerned Scientists reported that between 2002 and 2009, the cost of building a typical nuclear plant rose from $2 billion to $9 billion. Part of this is due to the huge regulatory hurdles that one needs to overcome to build a plant, but a lot of it is simply because the risk tolerance of these plants is so low that they have to be built to incredibly robust standards—and rightly so. But even if we suddenly decided that cost was not a factor, the engineering and construction requirements of building enough plants to satisfy world demand would be incredibly taxing—and then you'd have to mine enough uranium to keep them all running for the foreseeable future.

But let's say that we ignore everything covered so far and decide that the dangers of climate change outweigh the risks of an accident or the other environmental or monetary costs associated with building nuclear plants. Nuclear plants do produce a lot of power with little CO_2 released, so could they still be worth pursuing? The problem is that even if we started now, these plants take a long time to build, especially if the appropriate safety and regulatory measures are in place. Years of expensive permitting, environmental assessments, and community support for these projects precede a 5-to-10-year construction project. This effort would cost billions upon billions of dollars, and we'd still be stuck with water-hungry power plants that produce waste we have to store and that have to be fed nuclear fuel for the lifespan of the reactors.

At this point you're probably thinking I'm a NIMBY (not in my back yard) environmentalist who hates everything. I am not. Everything I have said here has been reported in reputable publications, and taken together I believe it leads us to the conclusion that every traditional source of power generation that currently exists will exacerbate climate change or bears side effects that make it untenable, or unscalable. Coal and oil generation are simply not compatible with a carbon-neutral future. In the case of hydroelectric dams, there simply aren't enough rivers to meet our needs, the dams are expensive and CO_2-intensive to build and eventually need to be decommissioned, and they result in large emissions of methane from the headwaters. Nuclear plants are safer than a lot of environmentalists give them credit for, but when they do melt down the results are catastrophic. They are incredibly expensive (and only get more expensive over time) and are unsuitable for hostile regimes that nevertheless need functioning electrical grids—and that we should be encouraging to get off fossil fuels, just like everyone else. In light of this evidence, I believe that it would be a mistake to rely on a massive buildout of nuclear energy—as some ecologists have called for—to solve climate change.

All these centralized power sources are themselves vulnerable to climate change. Remember earlier in this chapter, when I mentioned that large volumes of water are required in fossil fuel and nuclear plants to make electricity? Both hydroelectric and thermoelectric plants need abundant water, and that's bad news, because climate change will make water a lot harder to come by. In fact, in a 2016 paper in *Nature Climate Change*, researchers estimated that more than two-thirds of the world's electricity production is located in regions whose water supplies will be threatened by climate change. The researchers did say that some of this problem could be offset by plants being more efficient with water, but the fact remains that this is just one more example of the folly of our continued dependence on traditional methods of electricity generation. We are seeing a version of this playing out in Venezuela right now, at the time of writing. The country is undergoing a national emergency, as a drought has reduced the water available to the El Guri hydroelectric dam, which supplies 70% of the country's electricity. As a result of this drought, Venezuelans have severe power shortages and have had to reduce the length of their work week.

No, I do not oppose everything. In fact, I am a tremendous supporter of an immediate and aggressive buildout of wind and solar electricity generation—to an extent that makes some of my conservation science colleagues uncomfortable. Legacy electricity should come primarily from existing hydro and nuclear plants, with fossil fuel–consuming electricity generation phased out as quickly as possible, starting with coal-fired power plants. I would be very much in favor of solar panels, literally in my backyard (or better yet, on my roof), as they provide the best way forward for decoupling fossil fuel consumption from electricity generation. This buildout should occur both at the utility scale and on the rooftops of privately owned households in any part of the world sunny enough to support it.

Wind and solar are the only renewable energy sources that are getting cheaper with time and that are able to meet stringent modern environmental standards for industrial-scale projects. In 2015, the Bloomberg New Energy Finance group published a report that demonstrated that between now and 2040, the vast majority of new generating capacity would be in renewables. Since 2008, solar energy generation has grown 30-fold in the United States. Currently, China is the world's biggest installer of renewable energy projects. Even many developing nations are turning to wind and solar as they build up and modernize their energy grids. This wouldn't have been conceivable in even recent history and stands in stark contrast to other forms of large-scale generation, which are only getting more expensive.

Pragmatists will point out that there are technological and cost challenges in implementing an all-renewable power grid. However, I frankly think that if we were designing a power grid from scratch, we would find it very nonpragmatic to lock ourselves into a finite fuel source that we can only produce with great detriment to human health and the environment. As a society, we have granted incredible leeway to energy producers. We have collectively decided that it's acceptable to bear the risk of (rare) nuclear meltdowns, which cost billions of dollars to clean up and impact entire populations. We have shown a willingness to destroy farmland and relocate communities to build dams and then to pay out of the public purse to decommission those dams when we're done with them. In some cases, we're even willing to nearly bankrupt ourselves to build these dams in the first place. And I'm not even going to touch on the destruction that we've allowed many oil companies to impose on our health and well-being. Imagine the money we'd save if we chose instead to develop technologies that did not bear these catastrophic risks.

Aside from cost (which is dropping quickly), the other major objection to full adoption of renewables is their intermittency—their dependency on available wind or sunlight. Part of the solution to this problem is distributed storage, with homes and utilities maintaining large backup batteries that can store energy as it is generated. However, in a 2016 study in *Nature Climate Change*, researchers from NOAA (National Oceanic and Atmospheric Administration) argued that what we also need to do is invest in an improved energy grid that is capable of moving electricity over long distances with little loss— something that has been described as an "Internet for energy." The argument here is that if your grid is big enough, then somewhere on that grid the sun will be shining and the wind will be blowing. Therefore, you can generate energy in a climate that is amenable (e.g., by having solar plants located in a desert) and transmit it to where it is needed. So improved distribution, in combination with storage, will be able to solve the intermittency issue. Not easy nor cheap but possible, and certainly a lot more desirable than bankrupting ourselves to build legacy energy projects. It's no longer sufficient to complain about how hard a transition to renewables will be. We need to just get on with it and do what we have to do to make it happen.

To be clear, there will be costs in shifting our energy generation systems away from fossil fuels. There will be a need up front to mine a tremendous amount of rare earth metals and other materials needed for these technologies. Some of these operations have environmental costs of their own. And as we rush to more intensively use these technologies, we must not lose sight of the need to protect biodiversity and promote social equality. I certainly do not mean to let renewable energy projects off the hook, and they should be subject to rigorous environmental assessment like any other industrial project. Nevertheless, we live in a world with tradeoffs, and mining associated with renewables—

while expensive—is at least being conducted in service of a broader goal of reducing the amount of planet-killing gases that our traditional energy sources tie us to over the long term.

We cannot allow a comfort with the status quo to prevent us from embracing this transition. The solution to the electricity-generation component of climate change is therefore clear: a mix of solar and wind energy, backed by wide-scale deployment of battery-based storage technology. In this context, other mega-projects should be considered distractions from that core mission. Public policy should reflect this.

Am I saying that we can get rid of fossil fuels altogether? Not at all. In fact, we should cherish our fossil fuels and not waste them by shoveling or pouring them into generation plants and fuel tanks. Fossil fuels can be turned into valuable materials. They can be put into rockets and used to get us and our satellites to space (satellite-based technology is critical to conservation science). They can be used to produce high-value chemicals and compounds that improve our collective quality of life. Importantly, they are critical to the production of metals that are going to be needed for the mass buildup of clean technology that will underpin our sustainable future. So we should keep extracting fossil fuels in the most environmentally sensitive way possible (and at far smaller volumes)—we just shouldn't shovel them into furnaces to keep the lights on.

As we reflect on the broad problem of electricity's contribution to climate change, let's remind ourselves that we actually have a lot of control over this as individuals. We can alter our behaviors to reduce the amount we consume, and we can certainly replace inefficient devices with those that consume less power. The best time to do this is when they reach the end of their life, but there are also cases (such as incandescent lightbulbs) where it makes sense to upgrade at almost any time. And you can refine your usage, for example, by using a smart thermostat, so that you're getting the most bang for your carbon buck.

In prioritizing where you put your investments into efficiency, start first with the things you use a lot. Look for the lights that are on all the time or to the heating system that keeps you warm every day. These are the places where you can have the biggest immediate impact on your carbon footprint. When it comes to electricity production, you have the ability to do this at your home. It takes money, but as time passes that investment will likely drop. Net metering makes it economically viable, particularly in sunny regions, to install solar panels on your home and add clean energy to your energy mix.

But the bottom line is that any government that approves coal-fired power plants is certainly violating any carbon code of conduct. We need to save the fossil fuels for other purposes and get to the point where we don't need to burn them for electricity. Solar and wind power, combined with distributed energy storage, are the solution, and the technology exists right here, right now, to absolutely slash the energy sector's contribution to our climate problem.

SUMMARY

➔ Generation of electricity is responsible for a major portion of humanity's carbon footprint. The footprint of the electricity that you use depends on two things—how it was *generated*, and how *efficient* you are in using it.

➔ Electricity generation makes up about a quarter of CO_2-equivalent GHGs. The biggest single source of energy consumption in the home is usually the **heating and cooling**. Consequently, this is where your action should be focused.

➔ The single best thing you can do to reduce consumption is to keep your house temperature as similar to the outside as is comfortable. If you live in a hot place, use less A/C. If you live in a cold place, use less heating. This represents **reduction**. Where possible, you can **replace** inefficient heating units with more efficient ones, such as heat pumps. And ensure your house is well insulated.

➡ You can **refine** your heating through the use of a smart thermostat. These can pay for themselves very quickly in energy savings.

➡ Outside of heating and cooling, replace energy-inefficient technologies with efficient ones. Replacing incandescent lightbulbs with LEDs is a great first step. Buying Energy Star appliances is another. And get in the habit of turning things off when not in use.

➡ *Collectively, these actions can dramatically reduce your carbon footprint.*

➡ There are only two truly renewable sources of electricity that are widely available and can be installed at home: **solar and wind**. Solar panels are generally the most cost-effective technology for making electricity at home.

➡ We will all have to make decisions about what types of energy projects we support. The best types of energy generation from a climate perspective are solar, wind, and geothermal.

➡ *Coal is an absolute disaster,* and our first priority should be to eliminate it altogether from the world's energy mix. Large hydroelectric projects are better but bear substantial ecological costs and produce large amounts of methane gas—on top of the huge carbon emissions produced when they are built. Nuclear power produces relatively low carbon emissions but has serious problems of its own (including high levels of GHGs associated with mining the raw materials consumed in nuclear reactions).

➡ *Wherever possible, you should generate your own renewables or purchase from providers that offer it.*

TRANSPORTATION

In the United States, 27% of the country's carbon emissions come from transportation. In this chapter, we're going to talk about how to take a bite out of that piece of the emissions pie, through changes in behavior and by adopting technologies that can lead us down a cleaner pathway. Just like we need to shift away from fossil fuels for our electricity generation, we need to find ways to get from point A to point B without burning these compounds.

When you think about the way our transportation system is set up, it's kind of astonishing. An extremely complicated and costly chain of events needs to occur to keep our fleets of vehicles moving. First, someone somewhere in the world drills deep into the Earth's crust and pumps out unrefined crude oil. Maybe it came from a desert or maybe from the ground beneath the Amazon rainforest. Maybe it came from a well deep on the bottom of the ocean. We may have bought it from a theocratic dictatorship. Or perhaps we extracted it from oil sands—it begins as a sticky mess of tar-like petroleum that has to go through a complex, energy-intensive process to make it usable in our cars.

After drilling, we are left with a material that is, by itself, mostly useless—it must be shipped (in pipelines or in vehicles that themselves burn fuel) to refineries. Refineries then process the material into something we can burn in our gas tanks. From a typical 42-gallon barrel of oil, 19 gallons become gasoline, 12 gallons become diesel, 4 gallons are turned into jet fuel, and the rest ends up as a variety of other compounds. Once this step is complete, we have to load the fuel onto other vehicles for distribution to service stations around the world.

By the time you're pumping gasoline into your car, it has gone through a Rube Goldbergesque series of steps, each one consuming energy and emitting greenhouse gases. But despite the rising attention to the global climate problem, most people only interact with fossil fuels directly at that last stage, when they're standing at the gas station. This is the most visceral, hands-on experience that most people have to connect them with fossil fuels and climate change. The price of gas, the smell of the vapors, the experience of paying at the pump—this is what people think of when they think about oil.

The type of car that people drive is often a badge of honor, whether it's a gas-guzzling Hummer or a fuel-sipping Prius. But this choice is more than a symbol. Transportation is a massive consumer of fossil fuels, especially in North America. Cars and trucks alone account for about one-fifth of the emissions of the United States. This is absolutely huge—just one sector in the United States emits more greenhouse gases than the economies of entire nations. At the same time, in many cities our infrastructure is set up so that we are hopelessly dependent on cars to get from point A to point B and back.

Personal vehicles are one of the few items we own that are obligate consumers of petroleum. No matter how many solar panels we install, how many LEDs we buy, or how many wind farms we build, our current transportation system requires that we drill for oil, turn it into gas or diesel, and pour it into our gas tanks to keep our vehicles lumbering down the road. And every time a petrol-powered car is built, we lock ourselves further into this transportation network. For the next 20 or so years, that vehicle will need to consume fossil fuels, and someone will need to provide them to the person who owns that vehicle. While any single trip with a car doesn't make a big carbon impact, when you total up your repetitive daily driving across the entire year, the cost is substantial. In one year, a typical car used for an everyday commute may emit more GHGs than many people in the developing world do for all their activities put together.

Let's begin this conversation by stating one thing clearly: The best thing you can do for your transportation-related carbon footprint is to not use a personal vehicle that consumes gasoline or diesel for your daily commute. Public transit, alternate transportation, or carpooling is critical for day-to-day reduction of one's carbon footprint. However, we can't assume everyone will make this switch, because currently the vast majority of people in the United States do their commutes in a personal vehicle. According to the 2009 US census, 76% of trips to and from work in the United States are undertaken in a personal vehicle, with a single person driving alone. This tells us two things. First, we desperately need to shift away from personal vehicle use in the United States. Second, we need to adopt strategies so that those single-occupancy vehicle trips have the lightest footprints possible. In the first part of this chapter, we're going to focus on commuting with cars. We'll look at ways to get your personal vehicle's footprint down and discuss in detail the issue of electric vehicles, or EVs—a technological development that stands to revolutionize our car-related carbon footprints. We'll close out the chapter by exploring what is really the better solution, which is to shift away from single-occupancy commutes in private vehicles.

GAS GUZZLING

It's pretty obvious that not all cars are alike when it comes to fossil fuel use. A giant Hummer requires more fuel to go the same distance as a Honda Civic. From a climate change perspective, our goal should be clear: get as far as possible using as little fossil fuel as we can. Our personal transportation footprint is determined by three factors—the type of vehicle we drive, what we use to fuel it, and the way we use that vehicle.

When we're choosing what vehicle to buy, we face a staggering array of options. There is a lot at stake with this choice—when we decide on a car to purchase, we sink a large proportion of our personal wealth into that one item. This ties most people financially to a single vehicle for many years at a time, so we have to make sure to get this right. So how do we choose what vehicle to drive to minimize our carbon footprint without breaking the bank?

The first thing to consider is that the most environmentally-friendly vehicle is often the one you already own. Building a car takes a tremendous amount of resources. It takes hundreds of thousands of liters of water, vast quantities and varieties of metals, plastics, and other materials, plus loads of energy to power the manufacturing process. All industrial materials ultimately come from the ground, so all those components had to be mined somewhere on Earth, processed, and shipped to market. Each step in that chain is powered by fossil fuels, and consequently each step causes more emissions.

Owing to the complexity of the supply chain, the precise figure for how much CO_2 it takes to make a new car is difficult to pin down. One analysis, published by the *Guardian* newspaper in 2010, suggested that for every 1,000 British pounds spent on a new vehicle, about 720 kg of CO_2 is emitted. This is just for assembling the vehicle and doesn't take into account supply-chain emissions or the emissions associated with the production of the materials themselves. But the precise number is not all that important. The principle is that buying a new car produces a big footprint—carbon and otherwise. You're often better off keeping your old car well maintained and on the road.

This makes sense economically as well. There's a concept called the "total cost of ownership" (TCO) that you can use to compare the true cost of owning different types of vehicles. The TCO is the amount of money that a car (or any item) actually costs to own over its lifespan, including more than just its purchase price. This takes into account the fact that you don't just buy a vehicle, you have to maintain it, fuel it, pay for insurance on it, and so on. In fact, the biggest single cost of owning a new vehicle is depreciation—when you sell the car after many years of use, you will only recoup a small fraction of its initial value. In other words, the biggest single chunk of the TCO is usually the difference between the price you pay to buy and the price you get when you sell the car to its next owner. And the sharpest decline in the value of a car occurs in the first few years of ownership. As soon as you buy that new car and drive it off the lot, several thousand dollars of value just evaporate into thin air. So if you want to buy a new car for environmental reasons, it will cost you, just as it does when you buy any new vehicle.

However, as vehicles age they can cost more to maintain. There is clearly a point where an older car is no longer practical to keep on the road, and determining when that point comes is up to you. Thinking carefully about when to actually ditch a vehicle in favor of a new(er) one is a critical step to keeping your transportation footprint down.

Almost every car on the road is powered by gasoline or diesel contained in a fuel tank. The engine that converts the chemical energy in the gasoline into the kinetic energy of a moving vehicle is called an internal combustion engine. We'll refer to these as ICE vehicles for convenience.

There is a staggering variety of ICE vehicles, and if your goal is to minimize your carbon footprint, the reality is that many of the newer ones are quite fuel efficient. But as we've been stating, efficiency alone is an insufficient goal. Even a fuel-efficient vehicle is a problem, as it locks into our transportation system another vehicle that will consume fossil fuels for the next 20 or so years across multiple owners. It means that infrastructure—refineries, pipelines, delivery trucks, drill rigs—must remain in place to provide fuel for the fleet. If we really want to solve the climate problem, we have to get away from the need for sustained investment in fossil fuel infrastructure. That means it's time to embrace alternative fuels. Broadly speaking, there are three options here: biofuels, hydrogen fuel cells, and electric vehicles.

Biofuels are produced through decomposition of biological material, often crops like corn or sugarcane. They don't completely solve the problem of greenhouse gas emissions since the fuel is still mixed with petroleum, and CO_2 is still emitted during combustion. However, a tank filled with biofuel uses less petroleum than if you were to fill it with gasoline. Many people don't realize this, but the vehicle that you own right now may already be able to use biofuels. Take a look at the back of your car to see if you have a logo that says FLEX-FUEL. Flex-fuel means that the car is equipped to take a mixture of biofuels, including ethanol, methanol, or E85 (which is 85% biofuel). You can literally pump this biofuel into the car right now, provided you can find a place to buy it. Sourcing this fuel is the hardest part of adopting it, especially if you live in North America. There are a few Internet search engines that enable you to find gas stations that sell E85 biofuel. In Brazil, this type of fuel is already commonplace, so it's very easy to find.

Hydrogen-powered vehicles are a second alternative, and they receive a fair bit of media attention. The problem is, despite years of promises, they remain basically nonexistent. Like ICE vehicles, hydrogen vehicles harness chemical energy stored in fuel in a tank to move the vehicle. However, instead of combusting petroleum, they use energy contained in pressurized hydrogen gas. This gas acts chemically with oxygen in a vessel called a fuel cell, releasing energy and water.

People have been talking about the potential for this technology for a long time. I remember back in 2007 when then-premier Gordon Campbell of British Columbia, Canada, and then-governor Arnold Schwarzenegger of California (who had his personal Hummer converted to use hydrogen fuel) announced the government's intention to create a "hydrogen highway" connecting the west coast of Canada and the United States. The idea was that a network of hydrogen fuel stations would provide the infrastructure necessary for early adopters to drive emissions-free to the 2010 Vancouver Olympic Games.

The promise never materialized, and it illustrates the first problem with adopting hydrogen fuel cells. Just like with gasoline or diesel, you need to develop a massive, extensive fueling infrastructure. There needs to be a network of stations that sell the fuel, which needs a massive capital investment to get the technology off the ground—and no one will buy it if the infrastructure is not in place, no matter how much they care about the environment. The second problem is that the hydrogen fuel is itself produced through an energy-intensive process. You have to put a lot of energy in to get hydrogen fuel out, and so the amount of chemical energy left over to propel the vehicle is reduced.

The final problem, of course, is that you need vehicles that can actually use the hydrogen fuel. While there are many demonstration vehicles that have been built around the world, including buses, trucks, and forklifts, there just isn't evidence that manufacturers are lining up to produce hydrogen-powered vehicles. The Toyota Mirai is an exception. It is a fuel-cell vehicle available only in Japan. Honda also built a fuel-cell vehicle, but it has since discontinued the model. Despite these exceptions, it seems highly unlikely that hydrogen fuel cells will go mainstream. Of course, innovation can always prove predictions wrong. Perhaps there will be a breakthrough that will change the equation, but for now, there does not seem to be a serious chance for fuel-cell vehicles to overtake ICE vehicles as a primary means of transportation.

The best possible technology that exists right now are electric vehicles, or EVs. This is where the real future—and even the present—lies.

EVs are cars that are powered primarily by electrical energy stored in batteries in the vehicle. Unlike ICE cars, EVs can be charged with electricity produced by any means—including solar, nuclear, wind, or even fossil fuels.

Before we get into some of the benefits of EVs, we should revisit how ICE cars work, as it will help us put EV technology into the proper context. ICE cars produce propulsion through a series of controlled explosions that occur within the engine, which move pistons that provide power for the vehicle. We don't often think about this, but the process of converting fossil fuels into movement is a very complex one—hundreds of components are needed to precisely control the amount of fuel that goes from the tank to the combustion chamber, and the failure of any one of these parts completely halts the process.

Revolutions per minute (RPMs) describe how many times the engine pumps the pistons in a minute, and ICE engines can only effectively produce torque (rotation energy that powers the drive shaft) over a relatively narrow range of RPMs. To overcome this limitation, ICE cars have transmissions that use a series of gears to transmit exactly the right amount of torque to the wheels, so the car can move across a range of speeds. For a long time, changing gears required manual shifting (manual transmissions), but many new cars use automatic transmissions, where the car does the shifting for you. Transmissions are extremely complex and when they break, they can cost thousands of dollars to replace. While it's easy to complain when our cars break down, it's frankly amazing that automakers have achieved the reliability that they have, given the complexity of the machines they're building.

The number of individually fabricated components that go into this whole medley is staggering. Where there are moving parts, there are fluids to keep the parts moving smoothly. ICE vehicles require engine oil, transmission fluid, power steering fluid, and engine coolant. And where there are fluids, there are spills. While any one car doesn't spill a lot of oil, the total impact of all these little spills from every car in the United States is substantial. In fact, according to the US National Research Council's 2003 book *Oil in the Sea III*, runoff from roadways carried more oil to the ocean than any other human activity. Much of this had to travel through freshwater rivers and streams to get there, causing all sorts of subtle damage as it oozed its way to the sea. So all these ICE vehicles aren't just polluting our air, they're polluting our water as well.

An EV is fundamentally different. The energy that moves an EV is stored within rechargeable batteries. Like any electrical device, it can be charged using energy generated through any process that creates electricity. Environmentally speaking, this flexibility is one of the single biggest advantages of EVs. It means that when you build an EV, you're not building a device that guarantees another 20-plus years of obligate fossil fuel consumption. Its existence provides one less incentive for maintaining the network of petroleum gas stations and their supporting infrastructure. Furthermore, unlike hydrogen fuel vehicles, the distribution network for electricity for EVs already exists, through the electrical grids that service our cities. Every power line is a fuel pipeline for EVs, and as a result we avoid the need to develop a whole new set of pipelines and other infrastructure to power these vehicles (though some investments are needed for EVs to go mainstream; more on this later).

Even if you live in a dirty power grid, EVs are almost always more environmentally friendly than ICE cars. EVs are far more efficient. Whereas only around 20% of the potential energy stored in a liter of gasoline actually goes to making an ICE car move, EVs convert about 60% of their stored energy into motion—and this number is sure to improve over time. Not only that, but EVs actually recharge themselves when they slow down, through a process called "regenerative braking." It's as though your gas tank fills itself up every time you slow your vehicle! In fact, the only case where EVs may put out more GHGs is if your power grid is 100% coal (and this is debatable), and even then there are so many other environmental benefits to EV technology that they are still a better choice. For example, EVs do not leak oil into rivers and streams, because they simply do not use this product.

Regenerative braking hints at another EV benefit: they cost a lot less to maintain. You rely less on friction brakes to slow the vehicle down, meaning those brakes last longer before needing repairs. In addition, there are fewer moving parts in EVs than in ICE vehicles. EVs have no transmission, meaning that you have one less expensive system to maintain. The reliability of electric powertrains is phenomenal and getting better every day. And the less maintenance, the smaller carbon footprint associated with replacement components.

Finally, it's worth restating that ICE vehicles will consume fossil fuels from the day they're built until the day they hit the scrap heap. EVs unfortunate enough to be currently powered by coal do not share this problem. Grids can get cleaner over time, and people can choose to purchase power from clean sources or even generate their own. In any case, this is a reason to advocate for cleaner power, not to avoid adopting EVs.

Buying an EV can seem a little more complicated than purchasing an ICE car, largely because of confusing or contradictory information given to potential customers. In 2015, the US National Academy of Sciences published a guidebook called *Overcoming Barriers to Deployment of Plug-In Electric Vehicles*. In it, they argued that a lack of education about EVs is a major barrier to people purchasing these vehicles. So let's spend some time tackling this topic head-on.

Before we begin, let's clear one thing up—traditional hybrid vehicles are not true EVs. Hybrids, like the standard Toyota Prius, employ both an ICE as well as an electric motor to make the vehicle move. From the consumer perspective, you can think of hybrids as really efficient ICE vehicles. Under the hood, they have technologies that make them consume less fuel, but they cannot run on electricity alone. They're certainly a step in the right direction and are great in comparison to traditional ICE vehicles, but they do not get us off fossil fuels altogether.

By contrast, there are three classes of EVs on the market right now that can be run without any fossil fuels at all. The first is a plug-in hybrid, like the Chevrolet Volt. Plug-in hybrids are similar to traditional hybrids, in that they have both a small gas-powered engine as well as an electric motor. However, unlike traditional hybrids, these vehicles can run on electricity alone, at least for a short distance. The first-generation Volt can go more than 60 km (37.3 miles) on a charge, after which the gas engine kicks in and it behaves like a traditional hybrid. These cars are excellent for city driving, but over longer distances they require fossil fuels for their gas engine. My wife and I recently purchased a used 2012 Volt to replace our 2002 Honda Accord and have so far managed to completely avoid using gasoline on our daily commute. It is a beautiful vehicle and a great stepping-stone into the world of EVs. Where we used to spend $200 per month on gasoline, we now spend $20 per month on electricity for the same amount of driving.

The second class of EVs are the mid-range battery electric vehicles (BEVs), like the Nissan Leaf. This is a car that can go about 150 km (93 miles) on a single charge, without a drop of gasoline. While that range doesn't sound like a lot, studies in the United States have shown that 95% of commuter driving could be done within the range of a Nissan Leaf. BEVs have no gas tank at all and consequently require zero fossil fuel to operate. The third class of EVs are long-range EVs, like the Tesla Model S and Model X. Currently, an EV has to be able to travel 200 miles, or about 320 km on a single charge, to be included in this class. The large-battery variants of the Model S and X have ranges of over 400 km. The upcoming Chevrolet Bolt and Tesla Model 3 are other examples of this class of BEV. Like the Leaf, these vehicles have no gas tanks and rely purely on electricity.

The Models S and X, in particular, are stunning vehicles. Batteries aside, they are among the safest vehicles on the road. By removing the need for traditional ICE components, engineers were able to achieve unprecedented safety ratings in crash tests. The Model S is so well built that it broke the machine that tests how much crushing force the vehicle can take. These vehicles have batteries stored in a compartment at the bottom of the vehicle, giving them an extremely low center of gravity. This makes the vehicles resistant to rollovers. And did I mention that they're among the fastest vehicles on the road?

So why doesn't everyone already drive an EV? Aside from the fact that until recently these vehicles have just not been on the market, there are some factors to include in your decision before you adopt this technology.

PURCHASE COST

The first challenge with current EVs is the initial cost of buying them. EVs targeted at the mass market are more expensive than an "equivalent" ICE car. For example, a Toyota Prius with plug-in capability is about $6,000 more than a normal Prius. A Chevrolet Volt is about $40k to buy new, as is a Nissan Leaf (although very affordable used EVs are available). Both those vehicles, on paper, are arguably comparable in features to ICE vehicles in the $20–30k range. When you get to the high-end long-range EVs, value gets a little murkier. The cheapest Tesla Model S is US$70,000, which is clearly more expensive than most family sedans. But this is not just an EV—it is also a luxury vehicle, with the best safety rating ever recorded, and with a wide range of internal gadgets and features that target the luxury car market. The fancier Model S vehicles have incredible acceleration that rivals rare and expensive supercars.

I would respond to the EV price issue three ways. First, the initial purchase price is not the full story. Rather, the total cost of ownership (TCO) must be considered. Electrical energy is far, far cheaper than gasoline, and its price is more stable. So the fuel for an EV will save you money right off the bat. Maintenance is cheaper too, particularly for BEVs like the Leaf. Compare its maintenance schedule with any ICE vehicle and you'll immediately see what I mean—there are far fewer things to maintain on the car, at far longer intervals, than on an ICE vehicle. It doesn't take much searching online to find intrepid people who have computed TCOs for their own electric vehicles. One of the most comprehensive can be found at http://www.teslacost.com/. Here, an author named Paul Lee decided to evaluate how the TCO of the Tesla Model S compares to a number of modest ICE vehicles like the Honda Odyssey minivan over an eight-year period. The findings of this analysis were striking. The Model S was only a couple thousand dollars more expensive than the minivan over eight years of ownership, and the Model S has an eight-year warranty, covering its maintenance costs completely for that period of time. So while an EV may cost more up front, savings over time can greatly offset this purchase price.

Second, there are new EV products coming down the pipeline that stand to drastically alter the cost-benefit equation. For example, Tesla's Model 3 is supposed to be priced at around $35,000, before tax incentives. Nissan's next-generation Leaf is reported to have a longer range. Even Chevrolet has some tricks up its sleeve, with the pure-electric Bolt that promises a 300-plus km range and the second-generation Volt filling the plug-in hybrid role. By the time you read this book, there will doubtless be more options. As competition flourishes, and as battery costs decrease, selection will only get better.

Third, there is a robust used market for EVs, and as with any vehicle, purchasing a used car is a great way to save money. Leafs of only a few years of age can be found for half their original purchase price, and heavily discounted Volts are available as well. Tesla Motors has set up a certified pre-owned website, where used Model S vehicles can be had at substantially reduced prices relative to new vehicles, and these come with four-year warranties.

Finally, we do have to remember that 50% of the world's emissions come from the wealthiest top 10%. If you find yourself in that 10%, you bear a particular responsibility to find ways to help in the struggle against climate change, and replacing an ICE vehicle with an EV is a great way to do that. It reduces your own footprint, while also helping to encourage the market to support these products.

EV RANGE

The second concern people have about EVs is "range anxiety," or the fear of running out of power without having a place to recharge. Keep in mind, this concept only refers to pure EVs, not plug-in hybrids (as they have a gas-powered ICE as backup).

Range anxiety is a product of three factors: the range of the car, availability of places to recharge the vehicle, and charging speed. The range of any EV depends on the size of its battery pack. But it is also determined by how it is driven. Just like with an ICE car, if you accelerate lightly and drive smoothly you can go farther on a single charge. In addition, EVs of the future may be upgradable as battery technology improves. For example, if you now own an 85 kWh Model S, you can purchase a 5 kWh upgrade and get it installed after-market by Tesla. I've never heard of an ICE vehicle having an upgradable gas tank.

But range would be less important if we could charge cars as quickly and easily as we fill them with gas. EVs can be charged anywhere there is an electrical outlet, and there are far more outlets than there are gas stations. However, typical outlets will charge the car very slowly, as they are not designed for transferring energy at the rates necessary to fill up a giant EV battery. Nevertheless, typical outlets are acceptable for overnight charges, and studies have shown that more than 90% of EV charging is done at home.

But what if you need to charge quickly, for instance, if you're on a long road trip? Here, things get a little more complicated. Charging an EV is like filling a bucket with water using a tap. The size of the bucket is the car battery, and the tap represents the speed of the charging station. Big batteries, like big buckets, can hold a lot, but their large capacity makes them take longer to fill. Big charging stations are like big taps—they can fill the bucket quickly. The Model S and X have huge batteries that hold a lot of energy, but if you try to charge them with a normal electrical outlet they take a very long time to fill. But unlike a tap and bucket, there is a third factor involved in charging, the car's onboard charger, which accepts power from the outlet or charging station and stores it in the car's battery. Cars with big batteries tend to have more advanced onboard chargers, giving them the ability to accept more charge from high-amperage charging stations.

If the charging station is the tap, and the battery is a bucket, then the onboard charger is like a garden hose, in that it limits the amount of water that can flow from the tap to the bucket.

For example, let's say you wanted to fully charge a first-generation Chevrolet Volt. Its usable battery capacity is about 9 kWh, and its onboard charger is capable of charging at 3.3 kW. If you plugged it into a public charging station capable of delivering 9.6 kW, you wouldn't be able to accept charge as quickly as the station offered it. You'd be stuck charging at 3.3 kW, and it would take a little less than three hours to fully charge. If you had a Nissan Leaf, with a 6.6 kW onboard charger, it would accept the same amount of charge in about 1.5 hours. But if you had a Model S, with an onboard charging capacity of 20 kW, it's a different story. It would be able to take the full 9.6 kW to charge its 85 kWh battery. To fully charge the model S, you'd need to leave it plugged in for almost nine hours. However, that vehicle could take you more than 400 km (249 miles) on a charge, as opposed to 60 km (37 miles) for the Volt and about 150 km (93 miles) for the Leaf. If you charged each vehicle for the same amount of time, from the same charger, the Tesla would take you the farthest because its onboard charger would be able to accept the greatest amount of electricity.

To overcome the problems with slow charging speed, Tesla has built a network of "Superchargers" across North America, to be used with their vehicles only. These charging stations deliver about 120 kW. This means that you can charge your vehicle almost completely in about half an hour and drive as far as you could in a typical ICE vehicle. Thirty minutes may sound like a long time to charge a car, but consider that you would only use this feature on road trips, and you need to take breaks while driving long distances anyway. As you can see, the question of how fast these cars can charge is a complex one, but the reality is that most driving done by the average person is relatively short range, and the driver returns home every night. Nightly charges, even at slow rates, can leave you with more than enough energy for your daily commutes.

For EV roadtrips, you will have to do a little planning. There are resources to find public charging stations while you're on the road, and by the time this book comes out, there will surely be more. Plugshare .com is one resource that shows public and private charging stations throughout North America. Tesla's cars have onboard software that locates the nearest superchargers for long-distance travel. Sun Country Highways offers a service in Canada as well. But certainly, this takes a little more homework for now than it does to find a gas station. The bottom line about range anxiety is that, at least for short-range EVs, it is a real concern. However, as EVs become more mainstream, charging stations will become increasingly common (we will discuss how to help accelerate this transition in chapter 9).

Networks of charging stations are just good public policy. Charging stations make far better neighbors than gas stations. No one wants to live next to a gas station—they stink, and they're magnets for traffic and petty crime. But the worst part about them is what happens when you close them down. Depending on municipal bylaws, you often have to leave the land vacant for five years after the closure of a gas station. That's because gas tanks leach toxic chemicals into the soil, and it takes years for them to vaporize and for the soil to become safe. Sometimes, leaving them alone isn't enough, and millions of dollars have to be spent to clean up these contaminated sites. It's madness, when you think about it. We take perfectly good plots of land right in the middle of cities and make them unusable for years so that we can have convenient access to fossil fuels. Again, the benefit of EVs is not just what the vehicles do for the carbon footprint of commuting; it's also how they stand to transform our entire urban transportation infrastructure in very positive ways.

BATTERY FAILURE

One final concern about EVs is what happens when their batteries fail. Currently, batteries are one of the single biggest costs associated with

EVs. An 85 kWh Tesla Model S battery pack is worth about $40,000 at the time of writing—certainly no minor issue if it were to fail outside of warranty. However, over time the cost of these packs will come down. The first Toyota Prius hybrids had battery packs worth about $10,000 to replace. Now, that figure is closer to $1,000. A similar decrease in battery cost may be expected with EV battery packs over time, making replacements more affordable.

There is also a lot you can do to improve the battery life of EVs. All batteries have a finite number of discharge-recharge cycles, and that number can be extended by taking good care of the battery. Just like we regularly change the oil of ICE vehicles, there are certain basic maintenance rules for EVs. One such rule is to keep the battery charged only to 80% for daily use—filling it all the way to 100% is fine for long-distance travel but wears the battery out more quickly. In addition, these batteries come with extremely generous warranties. Both the Model S and Leaf have eight-year battery warranties, and real-world data show that EV batteries are extremely reliable. GreenCar Reports.com reported that after 80,000 km (49,710 miles) of driving, the average Model S retained 94% of its initial capacity. Our 2012 Volt has 100,000 km (62,137 miles) on the odometer and drives like it is brand new.

The concept of people replacing ICE vehicles with EVs is absolutely game changing, particularly if they generate electricity at home as well. In some ways, it hearkens back to a very low-tech age. Hundreds of years ago, all dwellings were somewhat self-sufficient. Most people grew their own food, with a surplus to sell at the marketplace. Everyone could, at the very least, darn and repair clothing, and everyone had basic skills in woodworking and construction. Some people would become masters in these crafts, but labor associated with keeping the household running was shared throughout the populace.

The industrial revolution changed all that. Suddenly it became more efficient to centralize everything. Production shifted to factories. Food generation shifted to industrial farms. People began to specialize. As electricity shifted from a luxury to a necessity, it also shifted to a model where a central utility generated energy on a massive scale, which was then distributed through a large network and sold to customers. This is the basic logic behind most of our energy generation—you generate huge amounts of power and sell it at a profit, with the money going to a single corporate entity, which may or may not be government owned. Many of us are stuck within monopolies, where a single energy company controls all of the electricity production and distribution, making it extremely difficult to pressure them to clean up their act.

The climate has paid a serious price for this efficiency, and solving climate change depends on us breaking free of this model. EVs go hand in hand with the concept of decentralized generation of renewable energy, which we covered in chapter 4. In a world where people generate their own electricity, and that energy then powers their vehicle, we will be a lot less dependent on electricity companies to maintain our standard of living. I dare say that EVs represent a technology that is very much in line with the most conservative of ideologies.

In short, there are challenges to consider when adopting EVs, but I argue that they're temporary and will be overcome as the technology matures. In many cases, they can be overcome right now. Voting with your dollar on an EV purchase—even if it costs a little more up front—is an incredibly conspicuous way to show that you are serious about conservation. It is not an empty gesture. Imagine a world where the overwhelming majority of vehicles are EVs and every house has a solar roof or wind turbine to charge the car. Yes, we'll still have to mine materials for these vehicles, but we're mining anyway—including huge mining operations for fossil fuels that we currently burn throughout the lifespan of our vehicles. EVs would help disrupt the entire supply chain that is currently built into making us dependent on fossil fuels, not to mention the money that spills from that chain into the pockets of denialist groups that derail efforts to address climate change.

And let me be clear—EVs are not strictly a symbolic gesture. A 2016 analysis by the Bloomberg New Energy Finance group demonstrated that if growth rates in EV sales continue, this single consumer choice could reduce global oil demand by two million barrels a day by 2023. To put this number in context, the 2014 crash in oil prices was associated with an oversupply of about the same amount. EVs alone could have a huge impact on the economics of fossil fuels.

PUBLIC TRANSIT AND ALTERNATIVE TRANSPORTATION

As efficient and progressive as EV ownership is, there's still no substitute for adopting alternative transportation. As I mentioned previously, part of the reason to get away from ICE vehicles is to rely less on the infrastructure that supports those vehicles. Reducing the overall number of miles traveled by personal vehicles has a similar function. Even if you're driving the most efficient vehicle in the world, you're still taking up space on the road, using parking spaces, and contributing to wear and tear on the motorways. So it should be our goal to decrease the number of vehicles.

Let's start with cycling. The bicycle is one of the most elegant devices ever invented. It is an incredibly efficient means of transportation. About 99% of the mechanical energy you put into the pedal actually turns the wheels. The energy you're putting into it is, by definition, biofuel—it's the food that you eat anyway as part of your diet. It builds physical activity right into your work day. From an infrastructure perspective, bicycles need far less support than cars. Smooth roads, bike racks or bike lockers for storage, and a place to shower when you get to work are all that are needed to facilitate this activity.

Well, almost all. The biggest barrier to cycling is, in my opinion, safety. Cycling itself is not dangerous, but combining bikes and cars is a recipe for disaster. We need separated cycling lanes and bicycle-friendly roads that promote safety (we'll discuss the policy implications of this in chapter 9). This is particularly true in cities where the public is often very "anti-bicycle"—and like it or not, this is a widespread sentiment, particularly in much of North America. Plus, if you get in a cycling accident with a car, more often than not (at least in the United States and Canada) the police will side with the car driver.

You can protect yourself somewhat by knowing your rights and responsibilities. Taking a bicycle safety course is one important step, and they are offered through many local cycling shops. These aren't just for safety—there are often quirky local rules you need to know about to cycle in your community. Taking a class can reduce the learning curve and help you quickly learn local cycling laws.

Bike thievery is a problem too. In Vancouver, most of my friends who cycled regularly have had their bike stolen, and usually more than once. Ask your workplace if you can store your bicycle indoors or in a locked parking garage. Alternately, bike lockers provide another way to safely store a bicycle when not in use. But if none of these solutions are available, walking is always an option for short distance commutes. If you're required to wear dress shoes for work, just walk in a pair of sneakers and change when you get there.

If your commute is too strenuous, then options exist to get pedal-assisted bicycles that can make the commute quite a bit easier. Traditionally, these have taken the form of gas motors attached to the drive chain of the bicycle. These can run about $300 to $400 and can be retrofitted onto your existing bicycle. But if you want to go all the way—and you're reading a book on climate change, so I assume you'll be receptive to this—you can get electric-assist bicycles. These are a little pricier, costing around $3,000, but they are an excellent alternative.

Public transit is an obvious alternative, but its availability depends on where you live. Some cities have excellent public transit. Others don't have it at all, or it is so dysfunctional as to be basically useless. But if there is any way to work public transit into your routine, it is worth it. There has been a lot of research on health and driving, and the findings are clear: driving for your work commute is not a healthy activity. There's nothing worse for your personal stress level than a long drive to work. Blood pressure, blood sugar, back pain, sleeplessness, depression—the list goes on. Driving is something that is repetitive and aggravating, and if you make one slip-up you can kill yourself or someone else.

From an economic perspective, replacing a couple of car-based commutes every week cuts your vehicle's wear and tear down substantially. Even if you can't make it all the way to work on public transit, a combined commute, where you drive to a park-and-ride, for example, is another great way to reduce your driving load and its carbon footprint. If public transit is completely out of the question, then using your vehicle to carpool is another wonderful alternative. Putting even one other body in the vehicle basically cuts your per-capita carbon usage in half, no matter how inefficient the vehicle. Not only that, but depending on where you are, it may give you access to HOV (high-occupancy vehicle) lanes or preferential parking.

The real challenge with our transportation footprint is to be creative. There are almost always ways to reduce the amount of carbon needed to get us from point A to point B, especially when it comes to commuting for work. Because we repeat this activity five days a week, fifty weeks a year, improvements here will add up to substantial carbon savings over the long term.

SUMMARY

➜ We expend a great deal of GHGs for transportation. We also have a lot of control over how we burn fossil fuels for this type of consumption.

➜ It takes a lot of GHGs to make a vehicle. For this reason, unless you drive a terrible gas hog, *the most environmentally friendly vehicle is often the one you already own.*

➜ First, **reduce** your use of your existing vehicle. Carpooling with even one other person cuts your footprint in half. Carpooling with more people cuts it further. Switching to one car instead of two saves you money and dramatically reduces your personal carbon footprint.

➡ When it's time to **replace** your vehicle, replace it with an electric vehicle (EV). These are cars that require no fossil fuels to operate. EVs help you **refine** your transportation-related carbon use by getting you farther while using much less energy than ICE vehicles.

➡ EVs are cost effective. There are many myths about them, but the most pervasive may be that they pollute more than traditional cars due to pollution associated with energy generation. In reality, they are far more efficient than traditional cars and are part of a structural shift away from fossil fuel consumption.

➡ *Reducing your reliance on a personal vehicle is an excellent move.* Eliminating it altogether, through the use of public transit and cycling, is better still. Beyond the benefits from a GHG viewpoint, public transit, cycling, and walking are far healthier, and people who rely on them report massive increases in satisfaction and reductions in stress.

CHAPTER 6

ADOPTING A
LOW-CARBON DIET

When we talk about a carbon code of conduct, it's easy to focus on the flashy technological solutions to the climate emergency—and admittedly, that's mostly what we have done so far in part II of this book. Clean energy, EVs, and home batteries are all critical parts of the solution. But to focus on those alone would do a disservice to our goal of a sustainable future. We have to talk about what we're eating as well.

When you sit down at the table, the items on your plate were put there with petroleum. Every step of the process, from production, to delivery, to the energy you used to cook the food, requires the input of energy and consequently creates an output of greenhouse gases. Before we can tackle the carbon footprint of our diets, we need to get back to basics and understand what goes into the food that we eat.

It all starts with plants. Plants get their energy from the sun and convert it to glucose through photosynthesis. Plants essentially breathe in carbon dioxide and exhale oxygen. They pull carbon out of the air and convert it into the solid physical structure that makes up their stems, roots, and leaves. Plants also need certain nutrients that are found in the soil, which are absorbed through the root system and pumped throughout the plant.

One of these nutrients is nitrogen, which is referred to as a "limiting nutrient." In other words, it is the one element that there usually isn't enough of for plants to grow effectively. It is the bottleneck in the plant's growth process, so if you add it, the plant can grow bigger and faster. Phosphorus and potassium are next in line—once you have enough nitrogen, you often need to add them as well, to help the plants grow.

To grow our food crops, we shower the soil with fertilizers, which provide the plants with an abundance of these limiting nutrients. We use a massive volume of fertilizer around the world to keep ourselves fed. In fact, the Food and Agriculture Organization (FAO) of the United Nations estimated that in 2015, the world used about 190 million tonnes (209 million US tons) of fertilizer. Historically, we used to get nitrogen fertilizers through mining something called saltpeter. But by the early 1900s it became apparent that there wasn't enough saltpeter to support our agriculture needs. The solution to this dilemma was found when one of the most important chemical reactions in history was developed by two German chemists, Fritz Haber and Carl Bosch. In this reaction, nitrogen is yanked from the atmosphere and is strapped onto three hydrogen atoms. This produces NH_3, or ammonia, a key ingredient that is used to produce nitrogen fertilizers.

This process was almost magical. It literally pulled the core component of fertilizer right out of the air, making it cheaper and easier than ever before to produce our staple crops. But it came at a cost. The Haber-Bosch reaction only works at very high temperatures, making it an energy-intensive process. Producing one ton of nitrogen takes the energy equivalent to that stored in 1,400 L of diesel fuel. Most of the world's fertilizer is produced in China, which relies heavily on coal plants to produce electricity, meaning that the electrical energy to make fertilizer is often incredibly dirty (see chapter 4). To make matters worse, the hydrogen atoms that are the building blocks of the Haber-Bosch reaction are usually derived from fossil fuels as well (usually methane from natural gas). So you're emitting carbon to power the reaction, and you're also emitting GHGs to build the fertilizer itself.

Taken together, about 1% of humanity's carbon footprint comes from the manufacturing of fertilizers. That's 1% of our carbon due to just one chemical reaction, in just one industry. But agriculture's carbon footprint doesn't stop there. The way we use fertilizer is also exacerbating climate change.

When we use too much fertilizer, we're not just providing nutrients for the crops we want. We're also feeding all the microbes that live in the soil, and these little critters are voracious consumers. Microbes release gas as part of their metabolism—in other words, they fart. And unfortunately, microbes fart nitrous oxide (N_2O), a greenhouse gas that's about 300 times more potent than CO_2 over a 100-year period. Microbes produce this naturally, but the amount of gas that they emit depends on the amount of nitrogen present in the soil. When we supersaturate soil with nitrogen by adding too much fertilizer, we create the perfect conditions for microbes to gorge themselves and produce a bewildering volume of this greenhouse gas.

In 2014, researchers from Michigan State University examined this relationship closely, and what they found was alarming. They determined that there is an exponential relationship between the amount of fertilizer used and N_2O emitted from soil. That is, in certain soil conditions, if you double the amount of fertilizer you use, you can increase N_2O emissions four-fold. With modern industrial farming, overfertilization is a rampant problem, so the amount of excess gas being released as a result of the misuse of this resource is a serious issue. In places with nitrogen-poor soil (common in the developing world) it's harder to overfertilize, so this is primarily a problem with farms in wealthy countries.

Despite being such a major issue, fertilizer is only the tip of the iceberg when it comes to our diet-related carbon emissions. Modern agriculture also douses crops in pesticides, herbicides, and other chemicals. This isn't done for fun—these compounds work together to help us produce the incredible yields of nitrogen-hungry crops that give us the low food prices we all enjoy. We also have to refrigerate the food, package it, distribute it, and so on. Every link in this chain releases GHGs.

The IPCC estimates that 24% of global GHG emissions are attributable to a combination of agriculture, forestry, and other land use. Another research group, the Consultative Group on International Agricultural Research (CGIAR) calculated this amount differently and concluded that a full third of global emissions are tied to our food supply. No matter how you count it, a major chunk of global emissions are due specifically to what and how we eat.

In my opinion, this is the most misunderstood and under-addressed component of our personal carbon footprint. Most people focus on transportation and electricity because innovations there are sexy and exciting, but diet is a major determinant of the emissions that we are personally accountable for. For those of us in wealthy countries, our diets are particularly carbon intensive, so we have a responsibility to reduce this part of our footprint. We have a lot of control over our own diets. We can often choose what goes into our bodies, so there is a direct linkage between our choices and our footprint. But understanding how to make responsible diet decisions can be difficult and sometimes counterintuitive.

THE 100-MILE DIET

Diet is the one aspect of the carbon code where reduction isn't the goal, for most people. We're not talking here about reducing caloric intake—for carbon reasons anyway—and in much of the world, people would do better eating more nourishing calories. Rather, our goal should be replacement of carbon-intensive food with that associated with a smaller footprint of GHGs.

The first step to fixing our diets is to understand which components of our food supply chain are energy intensive. Contrary to what people often believe, this is more about choosing the right types of food rather than focusing on foods produced locally.

In 2007, *The 100-Mile Diet* was published. It had a tremendous impact on readers. In this book, the authors demonstrated how hard it is to sustain a diet eating only products produced within a 100-mile radius of Vancouver, British Columbia, the city in which they lived. Some items, like fish, certain fruits, and mushrooms, were readily available, but others, such as cooking oil, were impossible to source. Basic items that we find in any modern kitchen were completely inaccessible. The book did an excellent job of showing people how complex our global food supply really is. It sparked a movement, where people calling themselves "locavores" shifted to local-only diets. While the purpose of this was not strictly environmental, many people seemed excited about the potential of reducing their food-miles (the distance that food traveled from farm to plate), in part because it's assumed that growing something nearby is better for the planet.

Surprisingly, this is not always the case. According to the CGIAR, between 80% and 86% of agriculture's global greenhouse gas emissions are emitted during the production of food—not the distribution. The specific values vary a lot by country—less-industrialized countries spend relatively more on distribution—but nevertheless, production represents the biggest piece of the food-related emissions pie. In other words, reducing your diet to a 100-mile radius generally shaves off a small percentage of your dietary carbon footprint, while switching to things that have low footprints to produce makes a much bigger difference. There are benefits to a local diet, but reducing your carbon emissions is generally not one of them.

So our focus needs to be on two things. The first is what we are eating. The second is how we produce it. As individuals, we have a lot of control over the former. We have less control over the latter—but our choices and advocacy can help change the system.

THE CARBON FOOTPRINT OF FOOD

Some foods are inherently more energy-intensive to produce, no matter how innovative we get in making them. The first principle of applying the carbon code to your diet is simple: If you're serious about climate change, you must eat less meat.

In chapter 5, we talked about how ICE vehicles lose energy when they convert fossil fuels into movement of the vehicle. In other words, not all the energy stored in the fuel can be harnessed, and much of it is lost due to friction. The same thing is true with biological organisms. When a cow eats grass, it doesn't transfer all that energy into building more flesh. Much of it is lost to metabolism—the processes that keep the animal alive.

In ecology, we think about the energy in ecosystems as if it's a big pyramid. At the bottom, you have plants. By converting solar energy and raw materials into plant matter, they are producing the base ingredients that all animal life ultimately depends on. This is why scientists refer to plants as primary producers. They produce the biological materials that everything else eats. At the next step in the chain, you have primary consumers. These are plant-eaters, such as cows. They metabolize vegetation and turn those plants into flesh that we consume. Secondary consumers eat the primary consumers, and so on.

When energy is converted from plant into animal, a lot of that energy is lost. Scientists call this "ecological efficiency." The concept was first articulated by a young scientist called Raymond Lindeman back in 1942. Along with colleagues, he identified a pattern that has been summarized as the "10% rule." The 10% rule states that as energy moves from one step of a food chain to the next, only 10% is stored by the consumer as flesh. For example, if a cow eats 10 kilos of feed, you'd expect it to grow 1 kilo of flesh. Cows aren't factories for converting grass to flesh—they have to breathe, sustain a heartbeat, and power their muscles to move around. Roughly speaking, these bodily functions take up the other 9 kilos of feed energy. According to the 10% rule, across the animal kingdom, the proportion of energy that goes to growth versus everything else is pretty consistently about 10%.

Of course, the relationship is far more complicated than that. Some species are more efficient than others at absorbing energy from food, and some ecosystems as a whole retain more energy as it flows from link to link in the food chain. But as a whole, the 10% figure is a pretty good descriptor of the concept of how energy flows through ecosystems. This explains, in part, why there are always fewer predators than prey in ecosystems on land. It's why you can never have as many wolves as there are deer without the wolf population declining.

The 10% rule applies to our food, just like it applies to natural ecosystems. When we eat plants, we can think of ourselves as retaining 10% of the energy that was locked up in the plant. When we eat a plant-eater (e.g., a cow), we only retain 1% of the plant's energy (10% of 10%). If a predator such as a grizzly bear ate the cow, and then we ate the bear, we'd be down to 0.1% of the original energy. If another bear came and ate us, it would retain 0.01% of the plant's original energy—and so on. This is the principle that explains why we don't farm lions and tigers for large-scale food production, as it would be far less efficient.

Farmers have their own version of this concept, called the feed conversion ratio, or FCR for short. The FCR measures how much output you get (e.g., mass gained, for a beef cow, or milk produced, for a dairy cow) per unit energy you put in. FCRs depend on the species of animal that is farmed. For example, cattle have an FCR of between 5 and 20 kg (11.0 to 44 lbs) of feed per kg of beef. Chickens are far more efficient, with a ratio of 2 to 1. Pigs are around 3 to 1. These values are better than what we see in natural ecosystems because of a long history of selective breeding and agricultural innovation that has sought to maximize efficiency. However, in all cases, it's still more efficient to just eat the plants directly.

Farmers are always trying to find better feeds for the animals they raise. Their goal is to get those FCRs as low as possible. There's an economic case here—they want to buy as little feed as possible—but the environmental goal of being as efficient as possible holds as well. Nevertheless, the reality is that the carbon footprint of meat is massive. Half of all fertilizer on Earth goes directly to producing animal feed.

In tropical countries such as Brazil, huge swaths of old-growth rain-forest are cleared in support of cattle ranching. In fact, clearing land for this single purpose accounts for 80% of the deforestation that occurs in the Amazon basin. According to the FAO, producing 1 kg (2.2 lbs) of beef requires the use of between 150 and 250 square meters (1,615 to 2,691 sq ft) of land. By contrast, producing 1 kg of pork or chicken requires just over 50 square meters (538 sq ft).

The next problem is specific to meat derived from ruminants, such as cows, sheep, and goats. In their four-chambered stomachs, these animals have special bacteria called methanogens, which ferment the feed in their stomach and convert it into something that the animal can absorb as food energy. The byproduct of this process is methane, and when cows burp, they emit this greenhouse gas. (As an aside, people often joke about cow farts killing the planet. Actually, between 92 and 98% of GHG emissions from cows are emitted orally. So it's cow burps, not cow farts, that are harming the planet.)

A single dairy cow puts out annual GHG emissions equivalent to a midsized car. And there are about 1.5 billion cows on Earth. On a global scale, this source of emissions is a massive problem. These animals are GHG factories, taking in plant matter and pumping out methane day after day after day, 365 days a year. From this perspective, beef and milk are like the coal power plants of agriculture. If we can deal with these emissions, we will go a long way toward solving humanity's GHG problem. Lamb is even worse, although lamb doesn't have the same status as a dietary staple as beef does in North America—so I won't focus as intently on it.

While it is bad news that Earth's 1.5 billion cows emit so much methane, there is a silver lining to this story. While CO_2 persists in the atmosphere for hundreds of years, methane only persists for decades. That means that if we can reduce our global production of beef and dairy cattle, we will see a response in atmospheric methane content more quickly than we would if they were emitting gases that last longer. In other words, cutting down on beef and dairy can be effective, and the effects are reversible over a shorter time scale. Consequently, our code of conduct for carbon should require that people replace beef consumption with alternative protein sources.

In a 2008 study in the journal *Environmental Science and Technology*, two scientists compared the relative benefits of switching the types of foods we eat (especially replacing beef with nonbeef items) with moving to locally based diets. They found that if you replaced a weekly serving of beef with chicken or fish, you would do as much to reduce your food-related carbon footprint as you would by replacing your entire diet with local foods. This is a shocking, counterintuitive result. The reality is that our food distribution system has become highly efficient—but the biological constraints of farming ruminants is not something that can easily be overcome through innovation. As a carbon-conscious consumer, it's a lot harder to switch to a 100-mile diet than it is to replace beef consumption with other food, so these results suggest that relatively simple dietary choices can pack a big carbon punch.

The key here is that food is habitual and is something we eat three times a day, 365 days a year. Any improvement we make to our diet is multiplicative. It may not seem like much to replace one beef meal with chicken once per week, but this means making 52 changes per year. If you've made the big decision to be a part of the climate solution, then the question of whether to choose chicken or beef answers itself.

While we should be changing our food choices, farmers continue to work to refine their production of meat. Improved FCRs are one way to do this, and new formulas are constantly being invented and tested to try to reduce the amount of feed (and therefore energy) that goes into producing a pound of meat. For beef specifically, new chemicals designed to reduce methane output are being tested as well. For example, in a 2015 study, a chemical called 3NOP was added to cattle feed. This additive reduced the amount of methane emitted by cattle by 30%—a substantial, persistent reduction in emissions across the study. It was shown to work in beef and dairy cattle, as well as sheep.

BEYOND BEEF—CHICKEN, FISH, AND OTHER PROTEINS

So let's say you swear off beef and other ruminants—what else can you do to reduce your carbon footprint? It's better still to switch to a non-meat diet. But I understand that many of us aren't ready to make that switch. Assuming you're not forsaking meat altogether, the common alternatives are chicken, pork, fish, and other aquatic animals. Unlike ruminants, none of these species possess the bacteria that releases copious amounts of methane during digestion, meaning that their carbon footprints are constrained to feed production and other energy used to raise the animals.

Does going organic or selecting free-range meats help your carbon footprint? The jury is out on that one. On the one hand, organic or free range does not change the FCR—you still have to feed chickens 2 kg of feed to make 1 kg of chicken, and pigs still need 3 kg of food to make 1 kg of pork. However, producing organic feed (a prerequisite for meat to be considered organic) can make a big difference. Recall that half of Earth's fertilizer goes to animal feed and that overdosing on nitrogen fertilizers can trigger exponential increases in N_2O. Organic farms tend to use less fertilizer or no synthetic fertilizer at all. This means fewer microbe farts, translating into less GHGs from feed production. For most people, making the decision to reduce beef consumption is far more important than worrying about specifics of how your chicken or pork is raised.

Please do not interpret my ambivalence about the carbon footprint of organics as a rejection of its value. There are all sorts of other reasons to go organic, but for the purposes of this book, I want to stay focused on climate change, which is already a big topic. And as I said in chapter 3, we don't want to paralyze ourselves with information—we want to make the simple choices that are generally right.

The carbon footprint of seafood is complicated and interesting. Around the world, we have done a lot of damage to the marine environment through excessive fishing. Over the past century we've netted, hooked, speared, trapped, and ensnared a lot of the biggest fish in the sea. As a result, a great deal of conservation attention has been paid to reducing the impacts of fishing. Some countries, such as the United States, are doing a great job keeping fishing under control, but globally, the Food and Agriculture Organization of the United Nations reports that 31% of fisheries are considered overfished.

In an effort to reward fisheries for embracing sustainability, environmental groups came up with something called an eco-label, the most famous of which is the Marine Stewardship Council. In a nutshell, these labels are awarded to industries that catch fish sustainably, at levels that the fish population can withstand for the foreseeable future. But these labels rarely explicitly concern themselves with the carbon footprint associated with fishing, and that footprint is not insignificant. A 2005 study in the scientific journal *Ambio* demonstrated that for each ton of landed fish product, an average of 1.7 tons of CO_2 are emitted. In the year 2000, about 50 billion liters of fuel were burned to land 80 million tons of marine life. This single sector accounts for about 1% of global oil consumption.

There are three ways to bring down the carbon footprint of the fish in your diet. The first is to make sure you're only eating sustainably-caught seafood. The logic here is that sustainable stocks are more abundant, meaning that you don't have to burn as much fuel to catch them. At the absurd end of the spectrum are some stocks of bluefin tuna. Plagued by persistent overfishing, populations of this species are now targeted by fishing vessels operating in tandem with spotter planes and helicopters that call out fish locations to the skippers on the water. This produces an extremely high GHG footprint per fish caught (not to mention its irresponsibility from a conservation perspective).

The second decision is to eat seafood that uses less fuel-intensive types of fishing gear. There are many ways that marine life is harvested from the sea. Some of these methods require a lot of energy to work. For example, bottom trawls are huge nets towed on the seafloor that scoop up animals that live on or near the seabed. These produce a great deal of drag, and they are among the most fuel-intensive gear to operate. By contrast, traps are lightweight gear that are baited and left on the seafloor. Just like mousetraps, fish traps are baited to attract their prey through the promise of a tasty meal. When the fisherman hauls back the gear after letting it soak overnight, it will be full of crabs, shrimp, or other fish species. Traps aren't towed through the water, so they also tend to require less fuel to operate. Favoring fisheries with low carbon footprints means you'll have to ask your local fishmonger how the fish you are about to buy was caught—and it can be very difficult to assess as a consumer. But learning more about how and where your fish was caught is a good habit to get into anyway.

There have been early moves toward developing electric fishing vessels for use in nearshore fisheries. In Norway, a fishing vessel has been created that can operate on purely electrical power for a full fishing day—not enough for offshore use, but sufficient for fisheries that operate close to land. We're a long way from this type of vessel becoming mainstream, but it's certainly not impossible to imagine, especially with the rapid innovation that we're seeing in battery technology.

From a policy perspective, I also believe that it is important to not subsidize fuel costs for fisheries. The free market doesn't always benefit the environment, but in the case of fishing, the free market is sending us a loud, clear signal: when fisheries aren't economically sustainable, they often aren't environmentally sustainable either. Helping out fishers with their fuel bills is not productive, as it reduces the incentive to be efficient and enables people to continue to fish well past the point where fisheries are economically feasible.

The third way to reduce the carbon footprint of seafood you eat is to eat animals that are low on the food chain. Currently, about a third of the fish caught worldwide goes directly into animal feed. Most of these are small fish that exist near the base of the food web, such as anchovy, mackerel, and herring. These species are no less rich in nutrients than salmon or tuna. There is no good reason, aside from fashion and taste, that we shouldn't be consuming these fish directly, particularly when they are fished sustainably.

This idea is particularly important when we think about aquaculture. This form of food production is sometimes vilified, and a full discussion of its advantages and disadvantages is beyond the scope of this book. However, not all forms of aquaculture are alike. Many conservationists tend to equate aquaculture with Atlantic salmon farming in open net pens, which is environmentally problematic in a lot of cases. However, in reality, we farm a dazzling array of marine species, each of which has its own conservation pros and cons. From a carbon emissions perspective, all of them are better than beef, and eating low on the food web is your best bet. Tilapia is among the most efficient protein sources you can get, and it is raised as an aquaculture product all over the world.

I would be remiss if I did not raise the issue of shrimp farming in southeast Asia. This is one area where I will break my rule of focusing strictly on carbon emissions, because the human and environmental impacts of this practice are terrible. Mangroves—aquatic trees important for biodiversity—are often destroyed to make room for shrimp farms. Among other benefits, mangroves provide buffers against tsunamis, so when you cut them down for a shrimp farm you're leaving the human population extremely vulnerable. It has been reported that the 2004 tsunami in Myanmar, which killed more than two hundred thousand people, was so deadly in part because of the loss of mangroves in the region due to shrimp farming. So don't eat farmed shrimp, unless it has been endorsed by the Aquaculture Stewardship Council or another reputable eco-label.

INSECTS AND VEGETARIANISM

There is one more protein source, aside from plants, that stands to make a big difference in the global food supply. That source is insects.

Yes, I know, it sounds crazy to many people, but industrial insect production could represent a major source of healthy, sustainable protein in the near future—provided we can get past the "ick" factor. People around the world have been eating insects for a long time. In Southeast Asia, deep-fried crickets, grasshoppers, and other insects are commonly found for sale by streetside vendors. You don't have to clear land to rear insects, and they have a very efficient FCR. Depending on the type of insect and what they are fed, they can be high in protein, healthy fats, vitamins, and minerals.

Insects also excel in the usable portion of their biomass. Let's go back to our environmental villain, the cow. According to the FAO, only about 40% of a cow's biomass is consumable. The rest is tied up in bones, skin, guts, and other body parts that we don't eat. Pigs and poultry are slightly better—we eat about 55% of those. But crickets are a different story. About 80% of a cricket is digestible, meaning we waste a lot less when we use insects to feed ourselves.

Currently, only a very small proportion of global protein consumption comes from insects, but industrial-scale operations are popping up across the globe, looking to monetize this environmentally friendly trend. But there is one barrier to adopting insects as a food source. That barrier is simply cultural acceptance, because the idea of eating insects grosses many people out. But norms change, and the idea is not all that far fetched. After all, we happily eat lobsters and shrimp, and they're basically just oversized bugs that live in the sea! I predict that as climate change raises food prices, and as cattle farming becomes non-economical (for example, by having carbon taxes levied that reflect the true environmental cost of the activity), insect farming will become a far more popular option than it is today.

The biggest, most aggressive way to reduce your food-related carbon footprint is to go vegetarian or vegan. If you've made the decision to do this, I applaud you. I'd also suggest that if you have committed to this pathway, you needn't worry about the carbon footprint of one plant crop over another.

GHGs associated with crops are usually measured in tons per hectare (or per 100×100 m plot of crop), and the values are really quite low. In Canada, values for cereal crops range from 0.65 to 3.38 tons per hectare, depending on species and region. None of the values are particularly high, nor would it make much of a difference if you focused your diet on one particular plant species over another. Some require more fertilizer than others, but most are incredibly productive.

Compared with beef and even other meats, anything you're eating from a plant is far, far superior in terms of its carbon footprint. If you've taken the step to vegetarianism, you can embrace the freedom of eating whatever you want from the plant and fungus kingdoms. On average, a North American diet free from meat will have about half the carbon footprint of a meat-heavy diet. A critical point here is that if you're ever in a position of organizing an event, make sure to provide a vegetarian option. This will reduce the carbon footprint of your event, demonstrate to everyone that you're serious about climate change, and empower your fellow environmentalists to keep their footprint to an appropriate minimum.

GREEN FERTILIZER

Your personal carbon footprint is closely tied to what you eat, and reducing the amount of animal protein—especially beef and lamb—you consume should be priority #1. But at a broader scale, how could we make our food system less harmful to the climate?

One way would be to remove fossil fuels from the fertilizer production process. As mentioned earlier, the Haber-Bosch process requires a lot of energy to heat up the reaction vessel, in which the chemical reaction that produces ammonia actually takes place. As we saw in chapter 4, electrical energy can be made many different ways—including through solar and wind generation. Therefore, there is no obligate use of fossil fuels at this stage. Whether we switch our electrical grids to clean energy or attach distributed solar and wind generation to fertilizer factories, there is no reason that fertilizer plants have to use non-renewable energy to heat up the reaction container.

Fossil fuels may not be needed to provide hydrogen for fertilizer, either. Back when Haber and Bosch first discovered their reaction, they used electrolysis of water molecules to provide the critical hydrogen in the reaction. Now, teams of scientists are revisiting this idea to find ways to produce ammonia without needing to pull the hydrogen from fossil fuels. Researchers from George Washington University developed a prototype process that employs a complicated bubbling technique to use the hydrogen atoms from water in the reaction process. Another team, at the University of Minnesota, attached a small-scale pilot fertilizer plant to a 1.65 MW wind turbine, which electrolyzed water with wind energy.

These efforts are experimental and are only at the prototype phase. But they illustrate that innovation is possible. You could imagine a future where farmers make their own fertilizer at small community-run plants, powered by wind and solar energy. Incorporating advances in agricultural science, they would then use only the fertilizer they needed to raise their crops.

Food items should bear carbon labels. Right now, if you walk into a grocery store and pick up a food item, you will easily be able to find a full accounting of the calories, micronutrients, macronutrients, and ingredients on the side of the package. For fruits and vegetables, the information can be found with a quick Internet search. The logic is that informed customers will make better choices about their eating habits and will eat healthier food.

There should be a similar label on food items that clearly show their carbon footprints. This is the second policy innovation that could greatly benefit our food system. An NGO called the Carbon Trust launched a program in 2006 in the United Kingdom that established a voluntary labeling system indicating the life-cycle carbon footprint of packaged food items. There are a range of similar programs out there, but none have taken hold in North America to date.

As customers, we need to push for these labels. It is far too difficult as people seeking healthy food choices to make educated decisions on what foods to eat to keep our carbon footprint down. I expect that the biggest opponent to such labelling would be the beef industry. If people understood the negative impact their hamburger created, I bet a lot more people would consider beef to be a treat rather than a staple food item that is regularly consumed as a protein source.

THE CARBON CODE OF CONSUMPTION

Your personal carbon footprint depends largely on your diet. It's totally under your control, but it can also be confusing. I have no doubt that savvy readers will identify cases that do not match the arguments made in this chapter. There are examples of trap-based fisheries with very high carbon footprints and trawl fisheries with low footprints. There are places where beef is produced relatively efficiently, and there are places where food distribution is inefficient and consumes a greater proportion of GHGs.

However, the general principle behind your dietary footprint is extremely simple. The more you replace animal-based protein with plant-based material, the lower your footprint will be. As you shift away from ruminants (beef, lamb, etc.) to other animals, your footprint will drop as well. In fact, if you do only one thing as a result of reading this chapter, it should be to minimize the consumption of beef in your diet. The further you go down this road, the better your footprint will be.

In the future, more carbon-friendly food products will become available. Keep an open mind, and give them a try. This is an excellent, tangible way to reduce your carbon footprint—and you might even have fun doing it!

→ *Agriculture is a tremendous producer of GHGs.* It is one of the single greatest polluters on our planet, both in terms of its climate footprint, as well as its impacts on fresh water, habitat, and biodiversity.

→ Your diet directly affects climate change. Substantially reducing your food-related carbon footprint is simple.

→ **Replace** or eliminate beef from your diet. By all metrics, beef is a huge outlier in terms of GHGs emitted. *Beef is the coal of food.* Substantially reducing your beef consumption is the single most effective thing you can do to improve your diet-related carbon footprint. Replace beef with other, lesser GHG-intensive forms of protein.

→ **Replace** meat from other ruminants as well. That includes goats and sheep. Like beef, these animals emit methane during their digestive process and are very carbon-intensive.

→ **Replace** animal-based proteins in general with those from plants.

→ Seafood is often a carbon-friendly option; however, ensure that you eat only sustainable seafood. Selecting products with an eco-certification, such as from the Marine Stewardship Council, is one way to do this.

CHAPTER 7

LONG-RANGE TRAVEL

F light is a luxury. The vast majority of people on Earth never set foot on an airplane or do so very few times throughout their lives. If you fly regularly, then you are statistically an outlier. This is hard to wrap our heads around in North America and Europe, where flying is relatively common. But every time we travel by air, we are accountable for a large amount of carbon pollution. Climate justice demands that we examine this carefully.

In chapter 5, we looked at how to reduce the greenhouse gases we emit during our daily commute. For the average person, these trips make up most of the transportation footprint. For those of us in the top quarter of global wealth, the GHGs we emit from long-distance air travel can be far more damaging. We can't ignore this any longer. The carbon cost of travel is enormous and growing quickly.

Let's reflect on a typical New York–London flight in a Boeing 747, a giant metal tube that weighs more than 300 tonnes (331 US tons) fully loaded. This metal monster has to rocket itself to just under 300 km/h (186 mph) to lumber off the ground. Then the plane has to raise itself 12,000 meters (39,370 ft) off the ground, overcoming both gravity and the resistance of air, and maintain a flying speed of around 800 km/h. To do this, the plane burns jet fuel—a lot of it. Every second, the plane burns 4 liters (1.1 gals) of fuel, consuming about 12 liters for each kilometer it flies. But the plane does not just release CO_2. It also spews a mixture of nitrous oxide and ozone which, emitted high in the atmosphere, has a bigger effect than it would if it were released on the ground. Therefore, calculating the climate impact of flight requires multiplying the CO_2 planes emit by a conversion factor, just as we convert the global warming impact of methane into a common currency that we refer to as CO_2-equivalent, or CO_2e. The UK government multiplies aircraft emissions by 2 to assess their impact (i.e., if 10 kg [22 lbs] of CO_2 were released at altitude, then we'd assume its impact is the same as releasing 20 kg at sea level).

On your New York–London flight, your share of the plane's carbon footprint is equivalent to a little less than 2 tonnes (2.2 US tons) of CO_2, or 10% of the annual carbon footprint of an average North American. Globally, the average person's footprint is about 4 tonnes of CO_2 per year—meaning that if you took 2 round-trip flights, that activity alone would be equal to all the GHGs emitted by an average human in a year.

The climate impact of flight is further exacerbated by the contrails that extend behind planes in flight. Contrails are the long white cloud-like streaks that you see when an aircraft speeds across the sky. Those streaks are created when gaseous water vapor in the plane's exhaust freezes at altitude. Contrails trap heat energy that would otherwise reflect out into space and add to the warming impact of flight. This is another reason that you have to add a correction factor when assessing the impact of flying.

In addition to the fuel used to move the plane, a multitude of resources go into supporting the flight. This includes the carbon footprint of building the plane itself, the energy used by vehicles at airports, the resources spent maintaining runways and airport facilities, and so on. The full life-cycle carbon cost of flying is substantial. The Air Transport Action Group, a UK-based industry association, estimated that the airline industry accounts for about 2% of global CO_2 emissions. That may sound small, but given the tiny percentage of the world population that flies regularly, it represents a massive outlier in terms of carbon-intensive behaviors. In the United States, air travel is responsible for about 8% of the country's emissions, making it the second largest source of transportation-related emissions, after cars and trucks. There's no getting around it—the carbon code has to apply to long-distance travel as well.

Let's start by discussing business travel. I want you to take a moment to think about all the business travel that you did in the past year. If you didn't do any, then this doesn't apply to you. But if you ever stepped onto a plane on behalf of business, ask yourself the following questions: Why did you take each of those trips? What were the goals? How many things did you accomplish on each trip? Could you have accomplished those things in another way? The true objective of business travel is not usually to experience the joys of sightseeing. The underlying purpose is to meet someone—to network, to collaborate on a project, or to sell something. Traveling is a means to an end. It just so happens that over the years, we've come to think of it as the primary means.

So to get a handle on our business travel, we need to rethink the whole concept of why we get on planes for work. Rather than starting from a position that travel is essential and trying to reduce that, let's start from the position that travel is *not* necessary. Now, each time you take a trip, you have to prove that statement wrong, and convince yourself that it is essential to the conduct of your business to take that trip. The carbon code's rule of reduction applies directly to flight—the first step is to find ways to do less of it.

Once we accept that principle, then we have to look for alternatives. What can we replace that travel with, so we can still accomplish our goal of meeting and working with people? Nowadays, conferencing software is frankly amazing. Google Hangouts and Skype are my two favorites. They even let you share a desktop view, so you can see what the other person sees on their computer. I rarely use my telephone for long-distance calls because the teleconferencing experience is so much better with these two programs. If we're working on a joint document of some sort, Google Docs provides an excellent space for collaboration, as does Microsoft OneDrive and Dropbox. These programs change quickly, though, and by the time you read this there will probably be different ones available. I have found that these pieces of software are excellent replacements for what would have been a short trip. If you've got one meeting objective or need to quickly connect people from all around the world, this is the way to go and is far more efficient than flying.

Once you've considered the option of telecommuting and decided it won't do the job (remember, we're starting from the position that you *don't* need to fly), the next step is to figure out how to make that trip with the smallest footprint possible. When you're picking a means of travel, you're making a trade-off among cost, time to destination, and the carbon cost of the travel.

Depending on where you're going, there can be many alternatives to flying. You could take a train or a bus—both are excellent low-carbon options, particularly if the trains are electrically powered. If you're travelling with multiple people, then carpooling is generally more efficient than flying, but if you're on your own then taking a personal vehicle is probably not the best option (unless you're driving an EV). The general principle here is that low and slow tends to be better than high and fast.

If you're covering a middling distance—say a 4-to 5-hour drive—it's often not all that much faster to fly than it would be to drive. Even if you're flying over a short distance, you still have to go through the hassle of getting to the airport, checking in, hoping nothing delays the flight, passing through security checkpoints, and then gathering your bags at the other end. If you actually look at the door-to-door time spent when a short flight is used, it is often the case that it's not all that much faster than alternative means. And in these cases, taking the plane is a lot less efficient.

That said, as someone living on the island of Newfoundland, I understand that sometimes flying is the only practical way to get to your destination. If you're going a very long distance, crossing a large body of water, or dealing with a tight time frame, then you have to take a plane. Fortunately, there are things you can do to minimize the carbon footprint of your flight.

First, never pay for first class or business class. These seats take up more space on the plane, and therefore they drastically increase your personal carbon footprint. In a 2010 article in the *Guardian* newspaper, it was reported that business-class seats are responsible for a 50% larger GHG footprint than economy seats. If an airline offers you this upgrade for free, you can feel free to take it, because it's not inducing demand for the seat. But you shouldn't pay for it, as it encourages expansion of these areas.

Second, try to find flights with as few stopovers as possible. This is because the most fuel-intensive part of flying is the takeoff. Many planes consume as much as three times as much fuel per minute during takeoff as they do once they've reached cruising altitude. Every time a plane climbs, it has to do a lot of work to beat gravity and gain altitude, effectively converting the stored chemical energy in its fuel tank to potential energy (due to height off the ground) and acceleration. At cruising altitude the air is also less dense, meaning that the plane experiences less wind resistance relative to being near the ground, and so runs efficiently.

Third, apply the **two-for-one principle** on every flight. That is, refine your travel plans so that when you do decide to fly, accomplish at least two goals rather than just one on your trip. You want to refine your plans to get as much productivity as possible out of each kg of CO_2 emitted. For example, rather than flying for a single business meeting, line up a few other clients and meet with them as well. If you're going somewhere to deliver a lecture, give the talk to more than one group. If you're travelling to make a sale, try to line up meetings with other potential customers while you're on the road. If you're collaborating on a project, link up with another working group to discuss other items. Or even piggyback an entirely different activity onto your business trip— why not take a few days to do some vacationing in this new place or visit an old friend while you're there? This is the two-for-one principle: to always get at least two things done in one trip.

If you find yourself flying to the same place twice in a few days or weeks, then you really have to reexamine your travel schedule and try to find other ways to be more efficient. This is something that we in the conservation and science communities should pay particular attention to. Many academics and conservationists actually fly far more than the average person, and the problem only gets worse the further you progress in your career. As scientists, we get credit for speaking at conferences, and the farther away the better. If we are often invited to speak, it reflects our stature as scientists. We need to acknowledge that we have this problem, without condemning ourselves to the point where we avoid travel altogether. Anti-conservationists certainly don't think about it at all, and we want to make sure that we're not putting ourselves at such a disadvantage that we prevent our important work from getting done.

Nevertheless, all of us who seek to heal the planet have to lead by example, and this means trimming the number of flights we take and making sure we get the maximum bang for our carbon buck every time we get on a plane. After all, carbon emitted in support of conservation has the same effect on the atmosphere as carbon emitted from industrial pollution. So as a conservation scientist, the first question I ask myself before taking a business trip is, What conservation benefit will come from this traveling? This is hard to quantify, but if we're keeping this principle in mind as we make our decisions, it can help guide responsible behavior and keep us from being wasteful. It will also help us be taken more seriously when we lecture others on their own carbon usage.

OFFSETTING

Once you do take that flight, you can choose to rehabilitate the atmosphere by purchasing carbon offsets.

Offsetting is an important, if controversial, means of reducing carbon emissions. The logic behind offsets is simple; you purchase a fund that guarantees that GHGs will not be emitted elsewhere, thereby theoretically balancing out the net amount of GHGs emitted into the atmosphere. At the risk of oversimplifying, you're basically paying someone else not to pollute who would have polluted otherwise.

There are a few ways this can work. The worst but most well-known example is probably tree planting. While planting trees is important, it is not an effective offsetting scheme because it does nothing to promote a structural shift away from fossil fuels, which are the ultimate cause of climate change. At best, tree planting produces a temporary carbon sink, as the trees grow up and suck CO_2 from the air. However, that carbon can easily be reintroduced by chopping those trees down or by burning the forest. It can also trigger perverse incentives—for example, cases where landowners will bulldoze existing forest for the purpose of planting trees and selling associated offset credits. Therefore, most reputable agencies do not endorse tree planting as a true offset.

A better approach to offsetting is to fund projects that do something to create a structural reduction in GHG emissions—for example, by funding the construction of a wind energy plant in a developing country or by providing poor communities with efficient cookstoves to reduce the amount of wood they burn to heat meals. A given offset project only counts as a true offset if it meets four criteria. First, it has to be permanent. In other words, the project has to exist for an indeterminate amount of time, so that the effect of the offset isn't simply lost after a few years. Second, it has to be additional. This means that the project shouldn't have happened without the offsetting investment (otherwise you're not really paying someone to reduce GHG emissions, as it would have occurred anyway). Third, the offset cannot be overcredited, meaning that the amount of credits that you purchase should match the actual amount of greenhouse gas saved because of the project. Fourth, the project should avoid leakage. Leakage happens if the offset project causes emissions to simply be moved elsewhere. For example, if you pay for an area to be permanently protected, but this simply causes habitat destruction to be shifted to areas adjacent to the reserve, this would be considered leakage.

The additionality principle is an example of how social justice often intersects with climate justice. Many of the best-certified projects are also ones that benefit the world's poorest in some way. For instance, the aforementioned cookstove project not only reduces the amount of forest chopped down to support cooking, it also means that people will be exposed to less smoke while they cook, thereby improving public health. Investing in the wind farm helps the developing world by providing it with clean energy, so it can become self-sufficient. This is the gift that keeps giving and means they won't go down the path of developing coal-fired plants instead. Their economy benefits, and the climate is harmed less. Ultimately, keeping climate change under control will mean that we have to spend less money mitigating its worst effects. So all told, effective offsetting is truly a win-win.

While the principle of offsetting is sound, there are certainly examples of scams. According to a 2015 report by the Stockholm Environment Institute, the majority of offsetting programs managed under the Kyoto Protocol did not meet key tests of integrity. In that report, they argued that we'd have been far better off if countries had simply met their emissions targets, rather than purchasing disreputable offsets from other countries. Fortunately, there is an NGO that runs a program designed specifically to assess the integrity of offsetting projects. This is the Gold Standard program (http://www.goldstandard .org/). They use strict auditing criteria to assess projects. If you purchase a Gold Standard offset, you can trust that you've done something positive for your carbon footprint. Plan Vivo is another excellent system (http://www.planvivo.org/).

So offsetting is useful as a way to rehabilitate the carbon emissions that you produce when you fly. However, the merits of doing this lag behind reduction, replacement, and refinement—those should be your primary tools for reducing the carbon footprint of your travel.

ACCOUNTING FOR YOUR CARBON

The carbon outputs of air travel have been extensively studied. Fortunately, all this research means that we can easily track the amount of GHGs emitted due to each flight we take. The UK's Department for Environment, Food, and Rural Affairs published a guidebook in 2013 on how to calculate and report GHG emissions, and this guide underpins most of the reputable online carbon calculators.

To calculate the carbon footprint of your flight, all you need to know is your origin, your destination, and how many stopovers you make. Punch these figures into an online calculator, and you'll get your result. I like the calculator at www.offsetters.ca but there are many that work just as well. These calculators will spit out the number of tonnes of CO_2e and will factor in all the effects, mentioned earlier, that make aircraft emissions more "potent" than those emitted at sea level. Playing with those calculators demonstrates that each flight is associated with a substantial emission of CO_2. It also shows how selecting an appropriate itinerary makes a big difference in the size of your GHG footprint.

Let me give a personal example. I live in St. John's, Newfoundland, but my hometown is Vancouver, British Columbia. There is no direct flight between these two locations, and it is impractical to drive or take a train—plus there would need to be a boat involved, because Newfoundland is an island. In general, I have three options for a layover. I can fly from St. John's to Calgary, Alberta, Montreal, Quebec, or Toronto, Ontario. The worst option is to fly through Toronto, producing 2.00 tonnes (2.2 US tons) of CO_2e. Montreal is slightly better with 1.80, and Calgary is best, at 1.67 tonnes. Just by adjusting the city in which I stop over, I can drop the direct GHG cost of my flight by about 17%.

Flying is particularly serious for organizations with lots of employees that travel. If air travel transitions from something done very rarely to something done commonly, then care should be taken to reduce that impact. That's where you, as an individual, can play a key role.

The first thing people need to do when cutting back on any activity—whether it be drinking, overspending, overeating, or flying—is to keep careful track of the extent of that activity in your life. Tackling a flight budget requires a three-step process. First, come up with a system of keeping track of the number of flights that are being taken. That database should include the location and purpose of travel, as well as how many stopovers were made on that trip. From this, you can use an online calculator to identify the flight's footprint, and you add that to the spreadsheet. If you want to get a little fancier, there are downloadable Excel spreadsheets online that allow you to automate the process of calculating the carbon footprints of your flights.

Second, set a goal for how many flights you think you can achieve—for example, how many flights can we cut without negatively affecting our business? Be ambitious here. Remember that ultimately all of humanity needs to mostly stop using fossil fuels to achieve climate stability. Therefore, just like with your personal travel, your default assumption should be that no flights are necessary, and then from there you make the case for each flight you take.

I'm wording it that way for a reason. The default opinion most of us have about carbon usage is that it's a given—we HAVE to fly, or our world would fall apart. When you look at it that way, you're never going to be very ambitious with your reduction targets because you're seeing it as taking away from something that's good. We have to start with an acknowledgement that flight is a necessary evil and is in fact not always necessary.

The third step is to make a plan for reaching your flight goal and to budget for appropriate offsets for that travel. Systematically go through each flight from the past year and figure out which could have been combined (through the two-for-one principle), replaced with an online meeting, or refined through better planning of stopovers and layovers. Of course, any business-class travel that was paid for should be replaced with economy travel.

These steps are particularly critical if you're an organization that considers itself an environmental leader. Universities and environmental NGOs should all have a carbon code of conduct, and it should include a regular review of flight activities. We have to lead from the front, not ignore this elephant in the room.

I am not calling for a complete abolition of flight by people who support the environment. Rather, I think our moral ability to fly to fight climate change can only be sustained by adopting institutional codes of conduct for carbon usage. Here, we can be transparent and open in our use of carbon, rather than beating around the bush with weak justifications for our behaviors. There is a time and a place for environmentalists to get on planes. However, if we don't make a serious review of our own activities, it will be difficult to convince others to reduce their own impacts.

VACATIONS

So far, we've focused mostly on business travel. But what about those cases where your travel has nothing to do with work, and you just want to go somewhere and relax? The carbon code won't have many fans if it doesn't allow for the occasional indulgence. In this section, we will investigate how to keep the atmosphere in mind when planning your next trip.

As with business travel, the single biggest source of carbon emissions on long-distance vacations is the flight to get there. Before we get started, let me just point out an excellent guide by the Union of Concerned Scientists. In 2008, they published *Getting There Green*, which was an incredibly detailed look at fine-scale decisions on how to keep the carbon footprint of a vacation to a minimum. That's the place to go if you want to get into details.

For our purposes, you can apply the carbon code to your vacation decisions. While I won't say you should reduce the number of vacations you're taking, there is a lot you can do within a travel plan to reduce the footprint of the trip. The first question is, How far do you want to go, and how fast do you need to get there? With respect to the former, the closer you can stay to home, the lower your impact will be. The fewer miles you have to travel, the less GHGs will be released in support of that travel. As with business travel, anything you can do to keep yourself out of a plane will generally be a good decision, from a climate change perspective.

Another factor here is the number of people you plan to bring with you on the trip. The more people traveling, the more unfavorable it becomes to fly. That's because when you calculate your carbon footprint of flight, that footprint assumes that there are other people on the plane with whom you are sharing the aircraft's total carbon impact. If a family of five buys a seat for each person on a plane, then their total carbon footprint is five times that of a single person. By contrast, if you put five people into a single vehicle, then your per-capita footprint is one-fifth of the vehicle. Taking a train or bus is better still.

There are all sorts of complexities in figuring out the precise amount of carbon emitted with each mode of travel. It's not as simple as train versus car. It depends how the train is powered or fueled, the fuel economy of your vehicle, the number of people on the trip, and so on. However, the distinction between ground transportation types is not as important as following the basic principle of filling up the vehicle with as many people as possible, using public transportation when it's available, and avoiding planes when you can.

When booking your trip, research the hotels and businesses that you plan to visit. Ask them how they are keeping their carbon footprint down (and maybe even ask to see their carbon code of conduct). The very act of asking demonstrates your interest as a customer, and the more people who ask, the more likely businesses will take steps to address their environmental impacts. While travel is carbon-intensive, at the very least you can make informed choices and support businesses that have made public commitments to minimizing their environmental effects.

But there is one type of vacation that is really hard to justify for any supporter of sustainability. I'm talking about ocean cruises. The Cruise Lines International Association estimates that a little over twenty million people take cruise-based vacations each year, so we have to discuss this.

As a marine biologist, I understand the draw of getting on a boat. Boats are fun. You get to explore, travel on the high seas, and enjoy being surrounded by the beauty of the oceans. But the industry as a whole has a very poor reputation in terms of its environmental impact. An NGO called Friends of the Earth (FOE) has done rigorous reporting on this topic and has demonstrated that the 16 biggest cruise line companies collectively released about one billion gallons of sewage in 2014. Most of this was untreated or only partially treated and was dumped directly into the ocean near the places they visited. On top of that, the ships emitted an additional 8 billion gallons of "gray water," water that comes from showers and sinks that is less polluted than sewage but is still dirty.

This being a climate change book, let's focus on GHG emissions. Cruise ships emit an incredible amount of air pollution. Like most large vessels, cruise liners are usually powered by bunker fuel, which is one of the dirtiest forms of fossil fuel. They emit large volumes of nitrogen oxides, sulfur oxides, and CO_2. It takes a lot of energy to move a floating city through water, and even if the boat is full of thousands of passengers, this translates to a lot of CO_2 per person. In addition, many ships also have diesel generators that are used to make electricity. This means that for many vessels, they continue to pollute even in port, when the vessel is not moving, as the boat has to burn diesel to keep the lights on.

Put simply, if you're selecting your vacation on carbon footprint, you wouldn't pick a cruise. They are quantitatively worse than other types of travel from a GHG perspective and produce many other environmental impacts that are beyond the scope of this book. This is even more true if you have to fly to the cruise ship terminal. There's just no way to justify that as a responsible choice from a GHG perspective.

I recognize that some people really love this form of vacation. If this is you, then you certainly have a lot of company. My first suggestion would be to try and replace a full-length cruise with something a little less carbon intensive but still on the ocean. For example, there are plenty of places in the world where you could stay at an all-inclusive hotel that offers whale-watching trips or something similar. In this way, you'd get the same vacation experience, including enjoying the marine environment, but with a much smaller footprint.

If you want to experience the ocean on your holiday (and you should!) and you don't require an all-inclusive resort, I strongly suggest eco-tourism. Speaking broadly, eco-tourism refers to a trip where the environment itself is the product. For example, scuba diving on a healthy coral reef is a great way to take in the beauty of the ocean while also voting with your dollar that the reef should be protected over the long term. Eco-tourism demonstrates that there is a tangible economic value to conservation. That's important, in a world where we're always fighting to make the economic case for conservation. And you certainly don't get that from a cruise.

If you're absolutely set on a cruise, then I'll refer you to the Friends of the Earth report card on cruise lines. Cruise ship companies are not all alike, nor are the ships they operate. In their report card, FOE assessed companies and ships on three dimensions: sewage treatment, air pollution reduction, and water quality compliance. They found that the Disney Cruise line was the best-performing line on the sewage treatment dimension. One of their vessels, the Disney Wonder, had the ability to connect to shore power when docked. This means it doesn't have to use diesel generators to produce electricity on shore, which therefore substantially reduces its air pollution when docked. Newer ships, with sewage treatment technologies, scrubbers to clean particulates out of exhaust, and other innovations, can have a far lower environmental impact than older ships. So if you're booking a cruise, consult the FOE report card, and ask the cruise line directly what it is doing to address its environmental effects and its GHG emissions. If you're a cruise fan, then I hope you engage with your cruise line about the issues I've raised in this book and get ready to buy lots of offsets to account for your trip.

SUMMARY

➔ *Flying is an elite activity that emits an incredible amount of GHGs,* which are particularly potent because they are released at high altitude. A low-carbon lifestyle requires that special attention be paid to this activity.

➔ Before taking any flight, whether it's for business or pleasure, ask yourself: *Do I truly need to take this flight?* Could it be **replaced** with another form of transportation, such as rail, or with no transportation at all (e.g., an electronic meeting)? Has the **two-for-one principle** been met (i.e., am I accomplishing a minimum of two things with this flight)?

➔ Take direct flights, and select the most direct route possible. Combine trips rather than making several separate round-trip flights. And never pay for business class.

➔ If you're flying in support of conservation, ask yourself: Is there a reasonable chance that taking this flight will result, directly or indirectly, in a net reduction in global carbon emissions? If not, is there a good chance that it will aid in developing resilience to the climate problem? At least one of these questions must be answered in the affirmative for the travel to be warranted—and even then, the carbon code of conduct applies.

➔ The carbon code applies to vacations as well. Apply all the above rules to your choice of flight.

→ *Cruises should be minimized,* but if you do decide to take one, investigate the environmental record of the cruise line and the specific vessel you intend to travel on. And for goodness' sake, don't fly to the cruise ship terminal.

PART III

SHARING THE CARBON CODE

In Part I, we reviewed the science behind climate change and faced its terrifying implications. I argued that we need a carbon code of conduct and that when we make decisions about using carbon, they should be based on reduction, replacement, refinement, and rehabilitation of carbon emissions. In part II, we covered the biggest sources of carbon pollution that we have control of in our individual lives and explored how the carbon code can help us reduce our personal carbon footprints in these areas.

Now, it's time to recruit others to the cause.

Taking personal responsibility for your carbon footprint means getting other people on board with the fight. This will be the focus of part III.

In chapter 8, we look at ways that you can constructively engage in the public debate about climate change and how a carbon code of conduct could work in the organizations to which you belong. I'll outline forums that you can contribute to, from the small and cost-free to the large and challenging, and highlight where you as an individual can help "shift the needle" in the public consensus on climate action.

In chapter 9, we identify actions that reputable experts mostly agree are things we should be doing right now to effectively build a carbon-neutral society. These are things that are bigger than us as individuals but that we should be pressuring our governments to implement. They are policies that would have clear benefits and that have been endorsed by a wide range of experts and qualified organizations. While any one of these actions will not be sufficient to solve the climate problem, they are all important pieces of the puzzle. Significantly, these are actions that are so sensible, you can always argue in favor of them and be confident that you're on the right side of the issue.

Then, we will wrap everything up with a recap of what we've discussed in this book. We'll revisit how tying everything together can help you be the protagonist in your own story against climate change.

CHAPTER 8

WINNING THE CONVERSATION

Over the past decade, former US vice president Al Gore has been a tremendous advocate for climate action. *An Inconvenient Truth* was a seminal piece of work that opened many people's eyes to the problem of climate change. He has a saying that perfectly describes one of the missions of the carbon code:

> I think the most important part of [stopping climate change] is winning the conversation.[1]

Climate change is a fact, and people who deny its existence and our responsibility for causing it are ignorant, misinformed, or lying. But when denial comes up in casual conversation, we have a choice to make. Do we allow their opinion to go unchallenged and avoid confrontation? Or do we take the chance to argue the case for climate action?

Mr. Gore picked his language carefully. To him, the struggle to get society to accept the realities of climate change is analogous to the ongoing efforts to stamp out racism, misogyny, and other social ills. Just like climate change, these issues are not solved solely at the upper echelons of government. Rather, they are tackled every day, with every conversation about the issue serving as a small struggle in a much larger conflict that will shape all of society. And it's an entirely appropriate analogy. The wealthiest 10% are putting out half of the world's carbon pollution. In doing this, we are committing violence against present and future generations, and that violence will be particularly harmful in the developing world. Climate justice is therefore tied closely to social justice, and denying climate change should be seen as an antisocial behavior.

As climate advocates, we have to get out of the mindset that the existence and cause of climate change is something open for debate. Climate change is real, dangerous, and caused by us, and I have yet to see any arguments against those facts that amount to anything more than misinterpretation or outright make-believe. We need to be unafraid to react with disgust when someone denies climate change. It's as ridiculous as believing in sorcery or a flat Earth. But how do we establish these norms without alienating people who see themselves as less personally invested in the climate struggle? How do we balance the need to win hearts and minds with the reality that we can't all be advocating for climate change all of the time, while acknowledging that a lot of people have more immediate problems in their lives? Herein lies the dilemma, and winning this conversation is at least as necessary as doing all the things for your lifestyle that we have outlined in the previous chapters.

FRAMING CLIMATE CHANGE

Imagine I offered you a million dollars to seal yourself in a tiny airtight closet for 24 hours. Sound like a deal? Of course not—you would die. No matter how badly you want that money, taking my offer is a nonstarter. You simply cannot survive without air because it is a physical constraint of being human.

Our global biosphere has physical constraints as well. If we exceed our global carbon budget, then our global temperature will go up by more than 2°C. Prospering in this climate-disrupted world will be very difficult. Even for those of us with wealth, we will have to devote a lot of our collective resources to moving out of flooded cities, fighting wildfires, rationing fresh water, and managing waves of climate refugees fleeing from wars sparked by drought.

In this context, let's revisit the way we talk about decisions as a society, particularly when it comes to projects that harm the environment. As it is now, our starting point in these discussions is almost always the same: We have to maximize economic benefits while minimizing impacts on the environment.

There is a major problem with that framing. It provides no upper boundary on the harm that can be done to "the environment." It implies that the default decision for an industrial project should be to do it, unless an overwhelming case can be made that it would damage the planet irreparably. It tips the scales in favor of harmful projects and often fails to consider anything other than local negative impacts. In addition, traditional framing implies that environmental degradation is okay, but that we should try to limit the amount of damage done.

The carbon pollution that is harming our planet is the product of many carbon-emitting activities conducted by people all over the world. If we continue to look at projects in isolation without considering their cumulative impacts, including their consequences for climate change, then we are almost certainly going to keep making bad decisions. There is a fixed global constraint on how much CO_2 can be emitted before we guarantee a planetary-scale disaster. Every decision we make has to consider this fact, and it must shape the types of projects we embark on at all levels of organization.

When it comes to the way we make decisions, it's time to flip our collective thinking. In conservation science, we call this the "precautionary principle." The precautionary principle states that we should select actions that are least harmful to the environment, even if we don't have a complete understanding of the harms that could be caused by that activity. It recognizes that "innocent until proven guilty"— which is a very ethical principle in criminal law—is not an appropriate standard for making environmental decisions. The precautionary principle is based on the fact that bad environmental decisions can be awfully permanent. If you cut down an old-growth forest, you can be pretty sure that you've done something very bad to the ecosystem, even if you don't have perfect data on exactly what was lost by destroying those trees. And that bad thing you've done can't be reversed within a human lifetime.

We need to be far more precautionary when it comes to the climate. The elimination of greenhouse gases from our economy has to become at least as important as making money over the short term. Because the negative effects of climate change are massive and will cost our economies dearly, it is no longer acceptable to make economic growth the main priority, with conservation happening on the margins. Instead, we have to draw a box around what environmental impacts are acceptable—and everything outside that box simply has to stop and be transitioned into clean alternatives. If the evidence is showing us that exceeding our global carbon budget will bring about a persistent living nightmare for much of humanity, it follows that it is of utmost importance to build a clean energy future, even if the cost is initially high. As technological barriers present themselves, they cannot be an excuse for inaction—rather, additional research should be triggered to overcome those problems.

I also acknowledge that what you're reading in this chapter falls well outside of the normal range of acceptable discourse that you'd see on the nightly news. People may say that it's too radical or unrealistic or that I as a marine biologist don't have the expertise to talk about it. But as a climate advocate, you don't need to have all the answers, nor do you need to fully understand the details of every aspect of the problem. You can do the right thing most of the time by following a carbon code of conduct based on the simple decisions outlined in part II.

Living by the carbon code means that you have to work past barriers and work to show others the merits of being part of the solution. When people challenge whether climate change exists or argue against evidence-based measures to fight it, then we have to find ways to challenge them back.

NOT ALL DENIALISM IS EQUAL

The most important message you can take away from this chapter is that **when you argue in favor of climate action, you are on the right side of history**. And while it may not feel this way now, a majority of people in much of the world would agree with you. In July 2015, the Pew Research Center conducted a worldwide survey to ask what people across the world thought were the biggest global threats. They found that in 19 of 40 nations surveyed, the public indicated that climate change was their single biggest concern. Across these countries, 46% of people were "very concerned" about climate change. For context, only 42% were very concerned about global economic instability, and 41% were very concerned about Daesh (also known as ISIL or ISIS).

In some parts of the world an even larger proportion of the survey population was very concerned. In Latin America, a median of 61% of respondents were very concerned about climate change, and in African countries the value was 59%. But in the United States and Europe, only 42% were very concerned. In the United States, politics plays a big role in determining whether people understand the threat of climate change: only 20% of US Republicans are very concerned about the issue, while 62% of US Democrats say the same thing.

The evidence is overwhelming that climate change is a serious threat, but clearly there are processes causing a lot of people to resist action. Getting into the breadth of the science of denialism is beyond the scope of this book, but we can unpack the basics of it here.

Denialism is on a spectrum rather than being an either-or issue. On one extreme, there are people who flatly deny that climate change exists. For these pure denialists, every argument about climate change starts with a complete rejection of scientific evidence. With these folks, arguing about climate action is pointless because they believe that nothing is happening. These are the people who ask: If climate change is real, then why is it cold in January? On the other end, we have people who fully accept the problem and advocate passionately for a widespread commitment to solving it.

For someone to act on climate change, they have to understand and accept three components of the problem. First, that climate change is real. The climate is changing, and it is changing rapidly. Second, that climate change is our fault—it is caused by the huge amounts of GHGs that we put into the atmosphere, mostly by burning fossil fuels, destroying forests, and farming carbon-intensive livestock. Third, that climate change is incredibly dangerous, and the overall climatic shifts that have already started to occur are overwhelmingly negative for humanity and much of the rest of the planet's biodiversity. But accepting the reality of climate change is not enough to catalyze action. To do that, we have to also convince people that action is possible, that it can be effective, and that the benefits of acting far outweigh the costs.

We can lump resistance to climate action into five non-exclusive categories of people: the benign denialists, the anti-conservationists, the corrupt, the not-my-faults, and the well-intentioned moderates.

Benign denialists are people who fail to accept one or more of the three climate facts but who do so simply out of ignorance (hence the "benign" label). They may deny climate change outright or accept that climate change is real but deny that humans are causing it. They may believe that it exists and that it's our fault but that it's not a cause for concern (i.e., they deny the evidence about its effects). But ultimately, these beliefs are based on a lack of knowledge about the problem. There is an entire industry devoted to obfuscating climate science and generating enough doubt to prevent people from seeking action. This makes it incredibly difficult to separate fact from fiction or evidence from nonsense. For people in this category, simple education may be sufficient to change attitudes. As an advocate, going on the attack against people like this can be counterproductive and can actually crystallize their denialist views.

By contrast, anti-conservationists work denialism into their personal identities. These are people for whom "not being a tree hugger" is part of their value system. They resist climate action and are proud of it—and they act this way even without being on a polluter's payroll. Anti-conservationists may actually accept climate science, but any concern for the issue is overshadowed by a "F___ you, I got mine" attitude. In this case, the individual may resist any action to address climate change, even if it were cost-free. They may even hold a mistaken interpretation of a religious belief, that the Earth is put here for human exploitation.

The category I call "corrupt" benefits financially from the status quo. These are the people for whom Upton Sinclair's famous quote applies: "It is difficult to get a man to understand something, when his salary depends upon his not understanding it." The corrupt are separate from anti-conservationists because they make money directly from denialism. Maybe they are paid to lobby against climate change, or maybe they run an industrial cattle ranch. In these cases they have a self-interested reason to resist climate action. In some ways, the corrupt are easier to understand than anti-conservationists because there is a clear motivation for their resistance against action.

Not-my-faults are people who don't dispute the facts around climate change but believe that any action should be taken by someone else to fix it. These are the folks who often employ the argument that we shouldn't do X, because Y country pollutes more than we do. They believe that their own impact, both in real terms and in terms of their ability to influence others, is too small to matter. As a result, they do not see a personal reason to get involved, nor do they believe in a moral obligation to cut their carbon footprint. Some environmentalists may even fall into this category, with the belief that they need not be accountable for their personal carbon footprints because they vote the right way or hold the right viewpoints.

The final category of resisters are the well-intentioned moderates. This category includes people who accept that climate change is real, our fault, and dangerous. They believe that action could be effective and even accept some personal responsibility. However, they make the argument that action is not cost effective. It would be nice to act, and we should, but gosh darn it, we just can't afford to do so, or doing something will just be too hard. We should embrace these moderates with open arms. They are one step away from being allies in the fight against climate change.

Why attempt to categorize beliefs? Because it is a common mistake among conservation advocates to assume that there is only one right way to push for change, and it is important to understand that resistance to change comes in many forms. Some people need to be educated. Some people need to be argued with. If you're dealing with an anti-conservationist, it might not even be about convincing them to change their mind but rather to make sound arguments so that other people watching your conflict are won over in the process. Far too often, environmentalists will argue that there is one true way to engage on a given issue, when in reality a diversity of approaches is necessary. What I am saying is, if you engage in any way that is comfortable to you, then you're probably taking a step in the right direction.

For the rest of this chapter, I'll cover some ways, big and small, to be part of the public debate on climate change.

POSITIVE ADVOCACY

One of the saddest papers I've ever read came out in 1999, in the *Journal of Personality and Social Psychology*. In this paper, two scientists asked study participants a simple question—how well do people's perceptions of their own competency line up with their actual competency? They asked this by having participants answer 20 questions selected from the Law School Admissions Test (LSAT), which were designed to assess "general logical reasoning ability." Before telling people their grade, the researchers asked everybody a second question: How well did you think you did, relative to other test takers?

As you might imagine, people's opinions often didn't match reality. The researchers found that the poorest performers grossly overestimated their own competence, while the best performers thought they didn't do as well as they actually did. The message was clear: incompetent people just don't see their own shortcomings.

The researchers concluded that the only way to make people aware of their own limitations was to increase their competence. But this creates a vicious cycle. How do you convince someone that they need to self-improve, when they already think they're smarter than you? How do you reach people who are actively resisting your message, as is often the case with climate change? How do you get past people's shields when they think they are smarter than the collective intelligence of the world's climate researchers?

The first place to start winning the conversation is by conspicuously living by the carbon code of conduct. Your actions will speak louder than words. But when you're ready to add your voice and try to change the conversation, the simplest, easiest, and friendliest way to enter into the public conversation is by engaging in **positive advocacy**.

In positive advocacy, you support action, rather than arguing against something harmful. This flips the script. Many people have a mental image of an environmentalist as someone who primarily opposes the development of large, impactful projects. There's a role for that, but we'll discuss it later. Positive advocacy might mean promoting EV charging stations in your community or supporting a carbon tax. You can do this by simply sharing interesting articles about clean energy on your Facebook, Twitter, or other social media outlets. You'd be surprised at how much your friends will read these feeds. There is nothing controversial about showcasing cool new technology that stands to improve the climate situation, and it's a safe way to get involved. Maybe someone will even get excited enough about it to go out and learn more. Ars Technica, Green Car Reports, and many other websites offer great examples of this type of article.

While you're getting excited online, don't be afraid to get excited in public as well. Take electric cars as an example. These are fascinating machines that are interesting, regardless of your political or ideological leaning. It's tons of fun to talk about these devices, and their adoption could make a real difference in our collective carbon footprint. Again, this is all part of setting norms. Many people think that trucks are cool and fuel-efficient vehicles are silly. There's no reason this couldn't be flipped, especially now that the best cars in the world are powered by electricity. Most positive advocacy does not require arguing at all—it involves getting excited about innovation. This is a great way to talk about the issue, especially with people who may fall into one of the categories of climate action resister.

A CARBON CODE OF CONDUCT AT WORK

We all work in teams, and teams have rules. Think about the teams you're a part of. Maybe you work at a university, or maybe you run a small business. Maybe you're the CEO of a Fortune 500 company, or maybe you're self-employed. Maybe you're a mayor of a city or governor of a state. In any of these cases, your activities at work affect the climate, and so your team should have rules about using carbon.

The core of the carbon code is to reduce, replace, refine, and rehabilitate carbon emissions. This works in your own life, as we've covered, but it is scalable. Your challenge is to figure out how to scale it to the organizations to which you belong. Try and help your workplace develop its own carbon code of conduct, and post it visibly.

Here's how to start. First, you're going to have to assemble climate supporters from within your organization. Find out if your employer has an environmental or sustainability committee. If it doesn't have one, talk to your boss about starting one. Many places already have these committees, but they are often inactive or lack leadership. Once that is done, brainstorm a big list of all the ways your organization uses carbon (I daresay getting them to read this book could be part of the effort). The pathway to cutting carbon will depend entirely on your organization's primary activities. For example, I work at a university. The best place for us to start is in looking at our air travel. Professors travel a lot, so we could start by recording how many flights our university employees are taking. From there, we will have a metric to assess ourselves so that we can set targets on reduction, replacement, refinement, and rehabilitation.

Next, come up with a list of actions that you could take for each of those high-carbon activities. For flights, I've suggested the two-for-one principle. This alone could cut flights by a half, but perhaps you can come up with other ideas around video conferencing or other forms of remote collaboration. Perhaps your company owns a lot of fleet vehicles, and transitioning to EVs would be a great way to reduce the carbon burden there. Ensuring low-GHG food is served in the cafeteria is another idea. It's really up to you and your committee to be comprehensive and creative. List the pros and cons of taking each action, and record why you ended up deciding (or not) to take an action.

Be transparent with these decisions. Write them down, post them online. Demonstrate to the world that your organization has a carbon code of conduct, and show how you have taken steps to follow it. Demonstrate the reductions in footprint you achieve, and broadcast them widely. Every business, every government, and every organization should have a carbon code of conduct, and you can be a part of making that happen.

Once you've taken care of your own backyard—you live by a carbon code, and your business or workplace has one as well—then there are ways to push the conversation even further. The first is to openly engage in public discourse through the written word and add your pro-climate voice to the media environment.

It is very important that individuals put pro-climate viewpoints into newspapers. To do this, you can write letters to the editor in response to articles that pertain to climate change. All it takes to get a letter published is to find the editor's email address at the paper. Then you just type up an email, sign your name to it, and send it in! It's as simple as that. Writing letters that are short and punchy is all it takes. If you don't consider yourself well-versed enough to write intelligently about climate change, don't worry: the columnist you'd be arguing with probably has no clue themselves. If you're arguing for climate action, you're on the right side of history.

Once you've cut your teeth on letters to the editor, it's time to take on something bigger. If you live in a democracy, then you have people whose job it is to represent you in government. So write them letters about climate change! Your letters can be topical, about an issue being debated at the time, or they can come completely out of the blue. Here's my challenge to you: find your representatives, whether they are local, provincial, territorial, state, or national, and send them something like the following:

Dear [Their name],

My name is [Your name here] and I am one of your constituents. As you're aware, climate change is a serious threat to our country's environment, security, and economy. I am writing to ask you to take this issue seriously and commit to reducing our country's greenhouse gas emissions quickly and aggressively.

Sincerely,
[Your name here]

You will be surprised how easy this is once you've done it the first time. And you will often get a response, because it's their job to do so. This may feel futile, but it does matter. In Canada, a member of Parliament (MP) is responsible for representing about one hundred thousand people at the federal level, but very few of these will take the time to actually contact their representative in government. Therefore, when an MP gets a letter, they operate on the assumption that many other people also hold the viewpoint presented in that letter. I've been told that when you write a politician, they assume that between fifty and one hundred other people share your viewpoint.

It's important to speak out, because politicians usually act when they know they will be rewarded in votes or campaign contributions. Think about gun control in the United States. This is an example where politicians are absolutely terrified to take action because the pro-gun lobby is so well organized. They know that if they make a move against guns, there will be a loud, immediate backlash among donors and at the ballot box. We need to get at least as organized as these groups and make it known that anti-climate decisions will be punished by the voters. Climate change affects everything, so when a politician acts against the interests of conservation, they need to know this will entail a heavy cost.

Petitions are another great tool to demonstrate collective will. Perhaps you want to support the creation of a carbon tax, promote the adoption of EV charging stations in your city, or even make your city's downtown core car-free (we'll cover some other things to advocate for in chapter 9). In any of these cases, your voice will be better heard if it reflects the will of many people. Attending an organized peaceful protest is another strategy. Protests are extremely effective at drawing attention to causes, and while governments don't like to admit it, there are plenty of examples of large protests causing social change throughout history. There are few things more satisfying than seeing a tangible change in your community and being able to say "I was part of that!" And you just might make some friends along the way.

RUN FOR OFFICE YOURSELF

If you're extremely committed to the climate cause, I'd like to plant an idea in your brain that you may never have considered: running for political office yourself. It's true, we need to hold governments to account and demand action on climate change, but imagine how much easier this would be if we actually had people in positions of power who consistently understood the scope of the problem and the many benefits to acting on it.

There are examples of politicians who have been reasonably supportive on the climate issue. Barack Obama has been a consistent advocate for climate action, and the EPA did a lot to promote climate-friendly development under his presidential administration. Former governor Arnold Schwarzenegger is another great example (and one that demonstrates that conservation can come from anywhere on the political spectrum)—he was tough on emissions standards in cars and remains a major proponent of reducing GHGs. California Governor Jerry Brown is similarly committed. By contrast, the intransigence of climate-denier leaders has caused tremendous damage to our ability to solve this issue.

Whether you're running for municipal, state or provincial, or national office, you will need to assemble the ingredients needed to make a run for public office. Let's take city politics as an example. Most city councilors are voted in by a tiny number of people. Turnout for these elections is often very low, so you actually don't need to convince all that many people to vote for you. You also may not need to align yourself with a political party at the local scale. By contrast, at the provincial, state, or national level, entering politics usually requires that you align yourself with a major party. In the United States, it would be hard to imagine running as a pro-climate candidate for the Republican Party, but there can be tremendous value in surrounding yourself with climate deniers and proudly, competently putting forward a green agenda. In Canada, there are many parties that hold reasonable views on climate change, and all could use your help.

If you don't want to throw your lot in with a major party, it can be a rewarding experience to run for a third party, even if they have little chance of victory. Green Parties are great examples of this. Their members rarely get into office, but when they do they can have a disproportionate influence on the media conversation about a given issue. Running as one of these representatives may give you access to town hall debates or other forums that can empower you to discuss sustainability in front of a mixed audience. Even if you don't win, you can shift the conversation.

Take all this advice with a grain of salt—I've never held office myself, so I can't speak personally to the challenges involved. And there are definitely challenges. It is a huge time commitment, and it can be very stressful. Your personal life will be on display for all to see, and the higher the office, the more you will be scrutinized. However, directing public funds toward the right types of programs requires that we either convince our leaders to take action or put people who are already convinced into those leadership positions. I feel that we environmentalists spend most of our time on the former and too little on the latter.

As a scientist, it is uncomfortable for me to advocate for political careers. Ideally, scientists should be apolitical. We're certainly allowed to have our personal opinions, but when our research is relevant to policy, it can be problematic to openly promote one candidate over another. But there is certainly a time and a place for doing so. In the 2015 Canadian election, scientists across the country came together in an unprecedented fashion for an "ABC" campaign, meaning Anyone But Conservative. We were just one of many groups that opposed the broad anti-science agenda of the previous government, but for us, this was personal. In countless ways, large and small, the government waged a war on science, particularly the type of climate science that empowers us to understand the scope and seriousness of the climate crisis.

In response to this pattern of hostility, many scientists took the unusual step of campaigning against the government. Those of us who could donated money. Others donated time, volunteering for campaigns of candidates who held pro-science views. Some people helped distribute campaign signs. Still others "phone banked," which refers to the process of soliciting people for donations and support via the telephone. Campaigns of all sizes are run by surprisingly few people. As a result, getting involved can have a big impact on policy, because it helps you develop a personal relationship with the person running. And if all that isn't reason enough, it's an excellent networking opportunity. Again, the pro-climate action of helping the right candidate win can also benefit you individually through the social benefits that come with volunteering.

Some people have chosen to push their climate action a little further with civil disobedience. Here, I'm referring to nonviolent action that disrupts the carbon economy and makes it difficult for environmentally destructive organizations to operate. I want to be clear that I am not advocating for breaking the law. I also remind you that civil disobedience can carry substantial legal ramifications—but depending on where you live, many countries have some amount of legal protection for lawful protests.

Now let me add a huge disclaimer here. I am only laying out what others have done and not advocating that you do anything specifically. And I definitely frown upon anything that crosses into violent action. Violent acts will do tremendous harm to the movement. The only way to properly assess the legal risks of participating in such actions is to hire a lawyer yourself and obtain legal advice.

But getting arrested in support of climate action is certainly noble. I'll give one example. In 2014, the American company Kinder Morgan began conducting exploratory work on Burnaby Mountain, British Columbia, which happens to be the location of Simon Fraser University (SFU), the university at which I received my training as a scientist. SFU is surrounded by a beautiful forest conservation area. Kinder Morgan had to do some assessments of the area to figure out what they would need to do to put a diluted bitumen (dilbit) pipeline in the park.

During these assessments, contractors came into the conservation area and cut some trees down during their survey work. They did not have permission to do this from the city, which manages the park, but they did claim to have federal permission. The locals were furious. Students and activists from the community set up a protest and physically blocked Kinder Morgan contractors from conducting the work. Lawsuits went back and forth between the city and the company. A court order to remove the protestors was ultimately secured. It was then that this little story, about an oil company doing a survey in a park, became national news.

The police set up a perimeter and told protestors that they would be arrested if they crossed the line. In response, community leaders from across the province arrived with the goal of defying the court order. One after another, scientists, community leaders, and aboriginal people willingly walked into the arms of police to be arrested. Many gave passionate speeches while doing so. Dr. Alejandro Frid, an ecologist whom I respect tremendously, said that he could no longer tolerate companies destroying our shared planet in the name of profits. Dr. Lynne Quarmby, the chair of the molecular biology department at SFU, was also arrested. These were not "professional protestors" or people who "should just go out and get a job." These were educated professionals who understood fully what was at stake.

The rage was palpable. While I wasn't there myself, I could feel it from my new home across the continent in St. John's, which I'd moved to only months before. How could an American oil company bully its way into our city—on a park no less—in the name of something that would only contribute to the hastening of devastation to our biosphere? By the time the crisis was over, more than 100 people had been arrested by the Royal Canadian Mounted Police. Kinder Morgan tried to launch lawsuits against some of the more outspoken perpetrators. However, it was later discovered that Kinder Morgan had given the wrong GPS coordinates when they secured their court injunction to keep protestors off the drill site; it turned out that many who were arrested may have technically done nothing wrong at all.

This story only had the legs that it did because of the courageous action of those who chose to commit to nonviolent civil disobedience. Such stories have played out again and again—climate marches that disrupt traffic, people arrested for blocking coal trains, and so on. These actions all play a part, and they all draw needed attention to this problem. Critically, they demonstrate to the powers that be that there are a substantial number of people who will not simply go along to get along.

ECO-INNOVATION

Over the course of this book we've discussed many examples of actions that can reduce one's climate footprint. However, innovation is still needed to make the transition away from fossil fuels happen quickly. We still need new clean energy technologies, and we need ways to make existing technologies cheaper and more efficient. We also need people to install them. This need represents a range of new careers that will open up that you, as an early adopter, are poised to participate in.

Our clean energy future will require armies of engineers, developing new technologies and improving on those that already exist. We will need people to design and build the power grids of the future. Botanists and agronomists will be needed to cultivate low-carbon agriculture, and ecologists will have to find ways to keep the biosphere as intact as possible during this grand transition. Economists and policymakers will develop the policies that governments can use to make these transitions cost effective. Lawyers will help draft environmental legislation, and political scientists will study how to make these laws politically feasible. Entrepreneurs will be particularly important as well. The market can be a powerful force, and new types of businesses will be part of the future's constellation of climate solutions. Put simply, it will be all hands on deck, as every profession reorients itself to contributing to the global climate struggle.

SUMMARY

➔ Climate change is deadly serious, but many people still don't understand the reality of it. *We have to "win the conversation" and help make climate-friendly attitudes go viral.*

➤ Defeating denial requires a cultural shift wherein denialism is socially unacceptable. People have many motivations for resisting climate action. Don't assume they're all villains—some are misinformed, while others don't understand the scale of the problem. Reach out to everyone.

➤ Engaging in the public discourse is an important part of living by the carbon code. **Don't let your personal lack of expertise on climate change stop you from speaking up.** If you argue for the climate, you are on the right side of history.

➤ The first thing you can do to win the conversation is to help your organization set up a carbon code of conduct. Start by agreeing on a code of conduct for your workplace, and post it for customers and employees to see. Identify your organization's most carbon-intensive activities, list potential ways to reduce, replace, refine, and rehabilitate, and then document the pros and cons of doing so. Set goals for your business and work to achieve them. Lead by example.

➤ Start with **positive advocacy**, where you argue publically *for* something that's pro-climate, rather than *against* an anti-climate activity.

➤ Once you've done this, there are many ways to engage in the public debate. These include posting on social media or writing letters to your local newspaper. Work with a political campaign and rally for pro-climate candidates. Run for office yourself, or even start an eco-business that explicitly serves to promote a shift away from fossil fuels.

➤ This list is not exhaustive. Be creative. Nearly any action here is better than no action at all.

CHAPTER 9

POLICIES FOR A
PRO-CLIMATE FUTURE

By this point in the book, you should be ready to make climate change personal. You've got your carbon footprint well in hand, and your workplace has a carbon code of conduct. You're convinced that it's time to win the conversation.

Now you just need something to converse about, beyond the actions you're taking in your daily life.

That's the focus of the present chapter. We're going to cover nine specific policies and solutions that should be put into action right away to advance the climate cause. Here, we're going to try and influence the governments that represent us, the companies we buy from, and the organizations to which we belong. I'm going to focus on policies where there is little or no gray area. These are things that experts around the world have called for, and for which the evidence is overwhelming that such a policy would absolutely benefit our collective effort to fight climate change.

Some of these would be expensive, and that's okay. Climate change is an existential threat to civilization, and mitigating it is imperative and urgent. We should not balk at spending to restructure our economies in a way that prevents the worst aspects of climate change, while simultaneously making the world a more livable, healthier, and overall better place. In any case, it's not as if this money disappears—it gets spent in constructive ways, on infrastructure and other things that create jobs along the way, thus stimulating the economy as we build a resilient society.

Unlike in previous chapters, the nine actions I will list are beyond your direct control to implement as an individual. These are the "big picture" actions that require collective work. These are the things that you can write your member of Congress, your senator, your member of Parliament, your president, or your prime minister about. You can have total faith that in arguing for these things, you are on the right side of history, because these ideas are backed by the best available evidence to date.

All the policies in this section are designed to shape our society in a way that keeps us within our global carbon budget. To be clear—even if all nine of these are implemented, they probably won't be enough. This struggle isn't over until our biosphere is intact, until we have halted deforestation and adopted sustainable agriculture, and until we have shifted completely away from fossil fuels as a mainstream energy source. But all of these ideas will help us take steps along that path.

Even if your government has agreed to one or more of these commitments, remember that a promise doesn't count unless action follows it. Whenever a government announces something progressive, there is always a cacophony of voices that will try and derail the process. So it's up to us as climate advocates to keep the pressure on and to let our representatives know that when they do the right thing, they will be rewarded at the ballot box.

1. MAKE POLLUTERS PAY BY ADOPTING A CARBON TAX OR CAP-AND-TRADE SYSTEM

Right now, the carbon footprint of things that we buy is rarely reflected in their cost. The price of a pound of beef versus a pound of chicken is totally disconnected from the emissions associated with either of those food sources. We know that it costs money to fix the damage caused by climate change—and we know that these costs are going to go way up, especially if we breach the 2-degree threshold. So who should pay?

A carbon tax is a way to make polluting costly and to ensure that the person or industry that does the polluting foots the bill. It is a tax based on the greenhouse gas emissions that are derived from burning fossil fuels. The principle behind a carbon tax is simple: carbon emissions cause demonstrable harm, and there is a financial cost to society to respond to that harm. If harmful goods cost the same as less-harmful items, then it's difficult for consumers to make informed decisions, and so even well-meaning people may buy items whose impacts we will eventually collectively have to pay to mitigate. Therefore, it is in society's collective interest to make polluters pay—even if those polluters are ourselves.

Carbon taxes are usually calculated as a price per tonne of CO_2 emitted, which is then reflected in the price of things that we buy. In my home province of British Columbia, the government implemented a carbon tax in 2008. It was the first jurisdiction in North America to do so, and to this day I'm very proud of them for doing it. This tax was paid at the point of purchasing fossil fuel. Carbon-intensive fuels, such as coal, had a higher tax, whereas natural gas had a lower price. As of 2012, the government computed a tax rate based on a fee of $30 per tonne of CO_2 emissions.

This was not just a tax grab. It was made to be revenue neutral because the government instituted various tax rebates, including an income tax cut, that offset the additional taxes taken in by the carbon tax. In fact, it was included in the law that if the tax was not revenue neutral, that the finance minister would have to take a 15% pay cut as a penalty!

As you may expect, this policy triggered howls of outrage by many people in the province. As you may not have expected, much of that outrage came from the left-leaning New Democratic Party at the time.

Opponents deployed all the counterarguments we've discussed in this book—it would make BC uncompetitive, it didn't matter because other people pollute more, and our contribution was so negligible that there's no point in punishing BC business! Impressively, it was the right-leaning government that actually implemented and stood by this tax.

Over eight years have passed since the tax came into effect, and the results speak for themselves. As of 2014, BC's fuel use was down 16% since the tax was introduced, while its economy grew faster than the rest of Canada's. The first year of the carbon tax was 2008, during the global economic meltdown, which caused its own drop in fuel consumption. That means that despite starting the tax when there was already a downturn, we nevertheless had a further decrease in consumption between 2008 and 2014. While we can't know what the economy's growth rate would have been without the tax, we can certainly conclude that the carbon tax was not an economic disaster for the province.

"Cap and trade" is another option to make polluters pay. In a cap-and-trade system, a hard upper limit (a cap) is set on the amount of emissions that are allowed within a given industrial sector, and all the players in that sector must trade credits to stay under the limit. If one group wants to pollute more, they have to purchase the right to do so from other groups, who then have to pollute less.

The United States uses a cap-and-trade system to regulate sulphur dioxide and nitrous oxide, two gases associated with acid rain in North America. Emission rates of these pollutants quickly dropped once the system was put in place, and the pH of rainfall recovered relatively quickly. In 2003, the US Environmental Protection Agency did an assessment of the success of this program—they found that the benefits of a cap-and-trade system exceeded costs by 40 to 1 for these pollutants.

In some places, cap-and-trade systems are already used to manage GHG outputs. European countries have been participating in such an arrangement since 2005. The state of California, as well as the province of Quebec, both have their own versions of these systems. Even the city of Tokyo has a cap-and-trade scheme in place.

Both the carbon tax and the cap-and-trade systems have their merits but are stymied by a lack of coordination at the national and international levels. It is admirable that small jurisdictions have taken it upon themselves to implement these policies—and if you live in a jurisdiction without one, then this is a great thing to push for. However, we also have to put pressure on national governments to implement large-scale carbon pricing systems that can demonstrably reduce greenhouse gas emissions. These agreements must be binding.

One final point: Some people disagree with carbon taxes from a social justice perspective, on the grounds that they could increase the cost of goods for the poor. My response to this is two-fold. First, increases in carbon tax should be paired with decreases in other forms of taxation, like the income tax rate for low- and middle-income earners. This would empower people to make decisions that reduce their overall tax burden by making smart carbon decisions. Second, carbon taxes should be enacted in conjunction with investments in public transit and other infrastructure that give people affordable alternative options to reduce their carbon footprints and therefore pay less carbon tax.

2. BUILD TOUGH REGULATIONS AND ELIMINATE CARBON OUTLIERS SUCH AS COAL

While carbon tax and cap-and-trade systems are good policy, they can be politically costly to implement. No one likes paying more taxes, and there are a million reasons someone could argue against implementing this type of policy. Therefore, an alternative is to push for very strong regulations for all industries that emit GHGs.

Governments are responsible for making rules. For most industrial sectors, there are rules and regulations about pollution that forbid the release of excessive toxic substances into the environment. One way that governments can successfully combat climate change with a lower political cost than a carbon tax or cap-and-trade scheme is by including GHGs as something that falls under government regulation. There are many examples of this being done successfully. California sets tight emissions standards for cars and trucks. In late 2015, the Obama administration released the Clean Power Plan, which set rules for GHG emissions in the electricity sector. Even airplanes are going to fall under such a scheme: in February 2016, the International Civil Aviation Organization announced a plan for all newly designed aircraft to meet a set of efficiency standards by 2020.

Sometimes these standards will have the de facto effect of making it impossible to build technologies that are GHG outliers. Coal-fired electricity plants are one example of something that we should get away from entirely. It cannot be stated enough how bad this source of electricity is for the environment and for public health. It's time for it to go, and we have no more excuses for keeping it around. If the cost of coal incorporated the damage it does to the environment, to the climate, and to our health, it would never be economical to use it in the first place. It poisons our food supply, spreads radiation, and contaminates the air that we breathe. It literally shortens lifespans in the regions where it is most heavily used. It made sense in the early 1900s, but it does not make sense now.

In 2008, prominent climate scientists Pushker Kharecha and James Hansen published a study in which they looked at the climate impacts of phasing out coal. They studied the forecasted concentration of CO_2 in the atmosphere by 2050 and compared what it would be if we did nothing to change our ways to a scenario where coal alone was entirely phased out by 2050. They found that in the business-as-usual scenario (i.e., the scenario where we don't take serious climate action), global CO_2 concentration would reach 563 ppm in 2100. By contrast, phasing out coal (as well as several other carbon-intensive energy sources, such as oil sands), would keep us at a peak of no higher than 446 ppm—still high, but within the safety limit to keep warming below 2 degrees.

In all cases, we should oppose the creation of infrastructure that supports the use of coal to generate electricity. This means opposing the mines that dig it up, the railcars that ship it, and the plants that burn it. It means opposing the use of our ports to ship it overseas, to countries that burn it in their own power plants. It means never building another power plant that burns coal, and it means making the cost of coal reflect the true cost that it imposes on the planet.

This will be bitterly opposed, not only by coal companies but also by people who see their livelihoods as dependent on the mining of this substance. I recognize that it's easy for me to say this, as someone with a job in a completely different field, but there is a strong argument to be made that many of these workers would be better off in the long term without these mines, provided that we as a society provide opportunities for coal workers to retrain in new fields. We can't condemn people to poverty in our march to climate justice—we have to make social justice a priority as well.

Shifting away from coal will take time, but it should be done as rapidly as possible. The longer we wait, the more urgent the shift will be—and if we continue to build new plants, then we're going to have to decommission these fully functional facilities before the end of their useful life. That's tremendously costly, so it's better to retire plants as they become obsolete. In the meantime, technologies that support carbon capture and storage should be developed and widely deployed. "Clean coal" technologies that reduce some of the more dangerous emissions from power plants should be a requirement, but should not be seen as an alternative to phasing out use of the substance. While we can't shut down these plants overnight, we can certainly prevent their expansion, effective immediately. No new coal plants should be built, and every incentive should be put in place to stop their spread. Speaking more generally, we need regulations that identify these outliers and eliminate them, replacing them with renewable sources. This should be a top priority in addressing climate change.

3. EXPLICITLY PRIORITIZE CLIMATE CHANGE IN DECISION MAKING

Most developed countries have some sort of approval process for large projects. Usually these involve an environmental assessment, a consultation with stakeholders, and a decision rendered by government about whether the project can proceed. Sometimes these decisions are made at the highest levels—for example, oil pipelines that cross international borders, such as the Keystone XL pipeline expansion, have to ultimately be personally approved by the president of the United States.

When assessing projects, many governments use an outdated view of environmental protection that considers all forms of economic benefits, from local to national, while environmental impacts are only considered in isolation on a localized scale. This may have made sense at one time, but it doesn't make sense in a world struggling with a global conservation problem. This paradigm biases decisions in favor of development, by allowing people to speculate on both direct and indirect economic benefits for any period of time they think is relevant, while simultaneously avoiding serious consideration of climate and other environmental impacts that will result from the project.

Let's think of this another way. When we talk about oil tankers, we often focus on what would happen if the tankers crash and cause an oil spill. In reality, even if the tankers were 100% safe, the ultimate use of the product is essentially a spill into our collective airshed that adds to the climate problem for the whole world.

Every major decision that we make should consider climate change as a factor. This should happen when we consider whether major infrastructure projects are in the public interest. It should also occur with more localized decisions—such as which type of vehicle a city or company will incorporate into its motor pool. The decision about whether and how to proceed with projects is important, because it locks in carbon emissions well into the future. There is a carbon cost of building but also a carbon cost of using that infrastructure, far into the future. We discussed this in part II—building a coal power plant ensures that for the next several decades, there is one more generator consuming coal and spewing pollution into the atmosphere, while also needing to be supported by fossil fuel infrastructure. If we wait until the plant has to be replaced due to its age, then building a clean replacement is cost effective. However, if we have to decommission the plant well before the end of its useful life, then we will have to pay a much higher price.

Explicitly including climate in decision making does not mean that all major projects would be halted. Instead, it would mean that an informed discussion would have to consider the global impacts of a project, rather than strictly a narrow assessment about whether the project is likely to cause localized harm. It also implies that a wider pool of stakeholders should be eligible to participate in a consultation process. If an oil sands project is built, it is not just relevant to the people who live nearby; it has implications for all citizens who are unable to opt out of climate change.

If you hear anyone talk about the environmental impact—positive or negative—of a project, two questions should always be asked: First, to what extent will this project promote an ongoing dependence on fossil fuels? Second, what is the total amount of greenhouse gas emissions that this project will produce over the lifetime of its use, and how does it fit into our global carbon budget?

If we seriously incorporate climate change into decision making, and if we respect the global carbon budget, then we're going to be stuck with keeping oil—a lot of it—in the ground. In 2015, two scientists from the Institute for Sustainable Resources at University College London published an analysis in the prestigious journal *Nature*, which demonstrated that two-thirds of known reserves must not be burned in order to stay under our global carbon budget by 2050. They showed that 80% of coal had to stay in the ground, along with a third of oil and half of gas. Nations such as Canada, with CO_2-heavy oil, can burn even less.

At small scales, such as within companies or municipal governments, a carbon code of conduct can be very useful to force the incorporation of climate change into decision making. Here again, when a group is faced with a carbon decision, they need only to ask themselves how they can reduce, replace, or refine their carbon usage—and if relevant, rehabilitate as well.

4. HALT ALL FOSSIL FUEL SUBSIDIES

One of the common arguments against climate action is that fossil fuels are cheaper than their sustainable alternatives. This is one of the ways that people justify constructing coal power plants, for example— that while they do have an environmental cost, the money you save on the cheaper plant allegedly makes it worth the cost.

Taxing carbon is one way to tip the scales in favor of clean energy. But the cost of carbon is further distorted by the fact that we have a vast network of government subsidies that promote the development and consumption of fossil fuels.

We shovel tax dollars into the petroleum industry like we shovel coal into furnaces. Some of these subsidies are direct, in the form of grants and investments given to firms doing research or exploration specific to the fossil fuel industry. Others may include targeted tax breaks or overly generous royalty regimes. Other subsidies are more indirect. For example, when oil is spilled, the polluter often gets away without paying for a full cleanup, and part of the cost ends up falling on the taxpayers.

In 2015, the International Monetary Fund conducted a study to figure out a defensible estimate of the total subsidies pumped into the fossil fuel industry each year. What they found was sickening. They argued that governments around the world collectively subsidized the fossil fuel industry to the tune of $5.3 trillion in 2015 alone. That's about $10 million per minute, every minute of every day, promoting a substance that if consumed unchecked, threatens to doom much of humanity and the other species with whom we share the planet.

I am not naive; even if the whole world adopted the carbon code, it would take time to transition away from fossil fuels. However, to slow the clean energy transition using economic incentives that benefit the status quo is collective insanity. As our global supply of easily accessible oil dwindles and we shift increasingly to fossil fuels that are hard to obtain (like those in the Arctic, those in oil sands, and fuels extracted through fracking), the subsidies will only have to increase to keep prices down. It's time to turn off the subsidy tap and at the very least let the free market decide which fuel is best (however, I think we should go further, as I'll argue in #5).

If we paid the true cost of fossil fuels, I suspect we'd be running as fast as we could to the nearest solar manufacturer. We'd be covering parking lots with photovoltaic panels, and we'd be developing wind farms in well-sited offshore areas. We'd be buying up batteries for off-grid systems like bottled water before a hurricane. We'd see bicycles and EVs in far more driveways, and our cities would develop around effective, affordable public transit. It would be comparatively far cheaper to adopt the clean energy lifestyle if we paid the true cost of carbon.

Once again, I am not alone in this call. In fact, in a 2015 meeting, the leaders of the G20 countries agreed to work toward a complete phase-out of fossil fuel subsidies. We need to keep public pressure on to ensure this happens, as leaders have made similar commitments in the past, without following through. This is one area where there should be a major overlap between economic libertarians and progressive environmentalists. If you believe that the invisible hand of the free market is always right, then we shouldn't be paying tax dollars to favor this industry. If you are an environmentalist, then you believe that we shouldn't be allowing polluters to get off the hook for the damage they do to our shared natural resources.

5. SUBSIDIZE CLEAN ENERGY

If we want to maintain our quality of life—and uplift the lives of those less fortunate—while seriously addressing climate change, the best option is to aggressively invest in clean energy and make it cheaper for private citizens to do the same.

Here, subsidies make total sense. It is in our collective interest to hasten the adoption of clean technology, and so using public funds to do this is good public policy. These subsidies should hit the full spectrum of clean energy and should include grants for research and development, low-interest or interest-free loans for reputable clean energy companies, and consumer incentives for adopting these systems.

It's pretty commonplace that clean energy projects get mocked in public media for receiving government "handouts." But the outrage is selective—we hand out money for all sorts of activities all the time. For instance, we've all agreed that it's okay to spend a tremendous amount on war. According to an analysis by the Congressional Research Service, the United States spent $1.1 trillion on the wars in Iraq and Afghanistan as of 2010. The world as a whole dumps about $1.78 trillion per year into its militaries. An additional $16.6 billion is spent by the United States on counterterrorism alone. Climate change stands to lead to more wars, so why not invest now and save money by avoiding a hypothetical war in the future?

What about something more benign, like our food system? We spend public money here, too. In 2010, the European Union spent about €39 billion on direct subsidies to agriculture. The United States, Canada, and many other countries also spend billions to support the way we farm. Subsidies to fisheries are similar—a 2013 report by the European Parliament estimated that the governments of the world provide more than $30 billion per year to fishing subsidies. We subsidize the energy that goes into our bellies, so why not the energy that could get us away from fossil fuels and keep our economies moving into a sustainable future?

Selective outrage around subsidies is particularly evident in the way we perceive the auto industry. Let's take 2009 as an example. In that year, the US government spent billions bailing out car companies. One fund in particular, the Advanced Technology Vehicle Manufacturing Program, was designed to spark innovation in clean energy. Tesla Motors was one of the beneficiaries of this program and was given a $465 million loan.

Some commenters flew into an apoplectic rage. The government was picking winners! How can we allow this? We have to let the free market work! Meanwhile, under that same program, the government gave $5.9 billion to Ford Motor Company. There was comparatively little outrage about that.

Fast-forward to the present. To date, Ford has yet to pay back the loan, and at the time of writing its only EV of consequence is the Focus EV—a car with a 100-mile range that hasn't sold more than a handful of units. By contrast, Tesla paid back its loan, with interest, by 2013, and put tens of thousands of EVs on the road. To date, it is Tesla that is still mocked for getting government assistance.

Now the purpose of this book isn't to weigh in on whether these other industrial subsidies are good or bad (except for fossil fuel—oil should not be subsidized). Rather, I'm highlighting that if we're willing to do it for so many other industries, is it really a stretch to extend assistance to the renewable energy sector? It's clear that society deems it okay for government to spend vast amounts of money when the conditions are right. If we accept that conditions are right to invest in clean energy, the question becomes: How much should the world subsidize clean energy?

As a thought experiment, let's use the amount of money we have already decided is acceptable to spend on fossil fuel subsidies around the world—$5.3 trillion per year. What could we buy for $5.3 trillion?

The cost of solar energy systems varies, so let's use the figure of $3,000 per kW. For $5.3 trillion, we could buy enough solar panels to generate 1.77 terawatts of energy. For reference, the US Energy Information Administration estimated that in 2010, humans consumed about 550 quadrillion BTUs of energy. This is roughly equal to 161,000 TWh, or the generation of 18.4 terawatts. In other words, for the amount that we subsidize fossil fuels in one year, we could theoretically build enough renewable energy capacity to account for 10% of humanity's energy needs. That estimate is conservative, because economies of scale would reduce the cost per kW of solar panels.

So if we spend an amount of money that we've already decided it's okay to spend, but spent it differently, we could have this whole mess sorted out in a decade. Now please understand—I fully comprehend that this wouldn't be as easy as I'm implying. You don't just walk to the store and buy enough renewable tech to power 10% of humanity. However, it's not an insurmountable challenge, and it is one that we simply have to meet.

What sorts of things could we subsidize? How about this: if you want to install a solar panel on your house, government should pick up two-thirds of the tab. Sound crazy? Not when you consider how much governments currently spend building new power plants. Imagine instead thousands of people employed in sustainable small- and medium-sized businesses, installing net-metered energy projects in tens of thousands of homes. The impact would be phenomenal, and the power would be completely clean.

Here's another one: Want an EV charging station at your house? We'll give you half off the sticker price. Buying a new car? Make EVs exempt from sales tax, and provide a credit at the point of sale. Selling your car in favor of a bicycle? You get a huge discount on your income tax (provided you give some proof that you're not driving to work anymore). Replacing incandescent bulbs with LEDs? A rebate of 90% at checkout.

In some jurisdictions, there are rebates in place to promote energy efficiency, but we need to ramp up spending in these programs. Rebates on energy-efficient lighting, insulation, and efficient heating and cooling are far cheaper than having to build new capacity. They will pay for themselves many times over and will reduce the amount of energy we have to collectively generate. It's already favorable to invest in clean energy—we need to make it economically insane not to.

In my opinion, there are two barriers to creative solutions like this being widely employed. The first is that our leaders are terrified that they will be slammed in the media for not being conservative with our funds. This will quickly get amplified by those who have a vested interest in our current system. The second is our collective lack of creativity. Some of these ideas are certainly "outside the box"—but we are dealing with a problem of unprecedented scale, which demands an immediate behavioral shift, along with sustained adoption of clean technologies.

So if you have a pen in hand and a piece of paper in front of you, write a letter to your representative giving them your permission to spend your tax dollars on clean energy. It's a common-sense investment, and it will benefit everybody.

6. DIVEST FROM THE FOSSIL FUEL INDUSTRY

For the 10% of us who are wealthy enough to produce 50% of the world's GHG emissions, we have to concern ourselves with investments. When we plan for retirement, we put money into stocks and other funds that grow over time. Big institutions do this too, investing in pension plans for their employees that can be worth billions of dollars.

Investing in a company is a major endorsement of that organization. You're basically saying that you support what they do and how they do it, that you want to see them grow, and that you trust them enough that you want to own a little piece of their business. If you support a transition away from fossil fuels, then it doesn't make sense to own fossil fuel companies. If you accept that GHGs are harmful, then it also stands to reason that profits from climate-harming activities are troublesome. Financially, legacy energy companies are not sensible investments—as we transition to clean energy, and as carbon taxes and other fees get imposed on fossil fuels, the industry will only become less profitable as its treasures have to be left in the ground.

Divestment is easy to propose but hard to do in practice. Most of us, if we own financial products at all, have money in mutual funds or other index funds, which own little pieces of many different industries. Fossil fuel companies are usually part of this mix, and we often don't get a say in precisely what is held within the fund. Our best bet to move toward divestment is to apply pressure on two fronts. First, pressure universities and institutions that manage our pension plans to divest. According to 350.org, a climate advocacy group, 503 major institutions around the world (with funds valued at $3.4 trillion) have committed to divestment over the next few years. Faith-based groups, pension funds, universities, and many other sectors have all participated in this effort.

Universities across Canada and the United States have growing divestment movements driven by students and faculty, and large schools such as the University of British Columbia and Harvard University are having to respond to these pressures. A 2013 survey by the First Affirmative Financial Network demonstrated that 77% of investment professionals expected growing risk in fossil fuel–based portfolios and reported that a majority of both individual and institutional investors were interested in divestment. And this risk is translating into a lot of lost money to institutions that have failed to divest. A 2015 study by *Corporate Knights* magazine argued that the University of Toronto has lost more than $550 million in its endowments as a result of failing to divest since the campaign to encourage them to do so started three years ago. Serious questions are going to have to be asked about whether universities are meeting their legal fiduciary duty to maximize investment returns when they fail to divest from an industry that will have to leave most of its assets in the ground.

A second way to advocate for change is to ask our financial institutions to create fossil fuel–free index funds for us to invest in. I am not an active day trader and prefer to invest money in low-fee index funds that deliver relatively stable returns over long periods of time. In doing so, I recognize that my assets are tied to some fossil fuel companies that may be mixed into my portfolio. This concerns me both because I'd prefer not to invest in these companies but also because I want to avoid exposing myself to the financial risk of investing in this industry. I'd rather buy index funds that exclude these companies. There are a handful of mutual funds and exchange-traded funds out there that favor climate-friendly businesses, but these are not mainstream at this point in time. Because they are not mainstream yet, they require that we as bank customers pressure our institutions to make them available.

Divestment does have some critics. Gerrit Heyns, the cofounder of a large investment management firm, wrote a *Guardian* article in 2015 that argued that most fossil fuel reserves are owned by state-run enterprises that would be unaffected by a divestment movement. He argues that divestment is a distraction, and we should instead focus on cutting emissions. My response is that he is totally right about the need to cut GHGs. However, we should be hitting this problem at every angle we can, and divestment movements have been effective at promoting change. The most cited success story is from the 1970s, when the apartheid government in South Africa was pressured by a large divestment movement. In addition, investment that does not go into fossil fuels will have to go somewhere else. Better that it go into funds that invest in exciting clean energy firms that offer far bigger rates of return and work toward solving the climate problem.

If governments start putting a price on carbon, cutting subsidies, and subsidizing clean energy, it is hard to imagine that fossil fuels would make for a sound investment—particularly if the scientific consensus is followed and most of the reserves are kept in the ground.

7. DEVELOP CYCLING INFRASTRUCTURE

In chapter 5, we discussed how cycling can benefit you as an individual. At the community scale, it can also play a major role in reducing the number of private vehicles on the road. In turn, fewer vehicles means less pollution to fuel climate change. From a carbon perspective, cycling is excellent. No fossil fuels are burned, and the majority of energy used goes right into moving the vehicle. Bicycles take very little in the way of resources to fabricate them, especially in comparison to cars—electric or otherwise.

So what is the rate-limiting step that stops people from adopting bikes? I'd argue it's the fact that going onto the roads of most North American cities on a bike is extremely dangerous. It's an easy argument to blame bad drivers (and there are certainly plenty of those), but the larger problem is that drivers and cyclists are sharing infrastructure designed to favor cars and ignore the needs of bicycles. What results is a hazardous mix that puts cyclists at risk and makes driving cars frustrating when bikes are around. We need infrastructure that separates the cyclist from the car. The best way to do that is with dedicated bike lanes.

This isn't rocket science. Biking will spread naturally when it is safe and pleasant and is not an exercise in danger and frustration. Cities should work to build interconnected cycling routes that are closed to cars and trucks. In practice, this can be hard to do politically. City councilors often have to spend tremendous political capital to make it happen. Amazingly, if they do get built, there are often citizens' groups that try to get them ripped up. One of the favorite arguments against bike lanes is that "nobody uses them." Oftentimes, this opinion is completely anecdotal and not backed by data.

Take the city of Vancouver, BC, as an example. There, the city invested several million dollars in separated bike lanes in the downtown core. These were intensely controversial, as they actually caused reductions in the number of drivable lanes in the roadway into the downtown core. As if on cue, the local talk radio stations were abuzz with angry people calling in to complain about empty bike lanes. "I've never seen a bicyclist in those lanes!" they'd bellow. Fortunately, the city had the foresight to collect hard data on the number of people who used the lanes. In the interests of transparency, they publish these numbers monthly, on a website, to demonstrate the full scale at which this infrastructure is used (http://vancouver.ca/files/cov/Bike-lane-stats-by -month.pdf). What it demonstrated was that hundreds of thousands of cycling trips were taken monthly across the city's various bike lanes—more in the summer and fewer in the winter.

Why is data collection so important in promoting cycling infrastructure? For one, bikes are very small in comparison to cars. Therefore, through the magic of confirmation bias, seeing a handful of bikes travelling down a bike lane can look an awful lot like no one using that lane. By contrast, if each of those people were barreling down the street in a Humvee, they would be very conspicuous, and it would look like more traffic. So even a highly used bike lane can look empty by the standards against which we judge normal traffic lanes. Only by systematically collecting real data can we assess how well the lanes are working. And if they are underused, then planners can work to make them more appealing.

But what about cost? It's true that separated bike lanes can cost millions of dollars, and perhaps more importantly, they can cut into the available road space for motor vehicles. However, if they can be used effectively to replace vehicle trips, then the city could save money on road maintenance, as bikes produce very little wear and tear on asphalt. In addition, it's excellent physical activity, and no doubt if more people cycled we'd see fewer cardiovascular and obesity problems—and therefore we'd save on healthcare too, over the longer term.

It's in your interest to have safe bike lanes even if you don't ride yourself. They get cyclists out of your way, so you can drive your car without worrying about them. They can actually improve traffic flow as well, since you're separating these two very different types of vehicles. Also, having more cyclists on the road means fewer people will need parking spots for cars, making it easier for you to park your vehicle.

So take any chance you can get to write your city council and tell them that you value cycling infrastructure—and if you're having an argument with a naysayer, remind them of all the great reasons to keep the cyclists on the road. Not only that, mention that you want to see programs to collect data on the cycling infrastructure, so its success can be measured and its design optimized.

8. PROMOTE EV INFRASTRUCTURE

At the time of writing, the major problem with EVs is not just that they cost more to purchase. The problem is that there aren't always many places to quickly charge them. Mid-range EVs such as the Nissan Leaf are primarily limited by range and the ability to recharge them on road trips. However, even a mid-range vehicle becomes far more capable if charging stations are widespread and easy to access.

The electrical grid is already like a network of pipelines for EV fuel. However, we still need places for cars to connect to those pipelines. This can be done through public charging stations or even just outlets made available to people who park their cars. All new parking garages and parking lots should be outfitted with several spots dedicated to EVs. A 2013 report by Deloitte estimated a typical installation cost of around $4,000 for a public charger, but it can vary widely. If a city or organization wants to install DC fast-charging stations, that's great as well—but those are costlier (around $50k-$60k).

While I am a proponent of making these stations free to use to encourage EV adoption, it is possible to recoup installation costs by charging users for energy. For example, Chargepoint is a company that supplies a wide range of charging stations, many of which employ a system for charging users that can be set up by the owner. If a condo or business wanted to provide charging services but pass the cost on to users, a Chargepoint device could allow them to do so.

Selling power through these stations can be far more profitable than operating a gas station. When people open up a gas station franchise, only about 1% of the price paid at the pump actually goes to the station's owner. The rest goes to the oil company, taxes, and other operating costs. By contrast, power sold through Chargepoint stations can be marked up many times the base price per kWh. A transaction fee goes to Chargepoint, and the owner earns the rest.

The responsibility for developing EV infrastructure is shared among governments, businesses, and private citizens. If this issue interests you, strike up a conversation at the businesses you frequent. Write in their suggestion box that you'd like to see charging stations in their parking lots, and talk to the employees about this suggestion. Social media is another great mechanism—publicly ask whether companies intend to install EV chargers, and praise them if they do. Point out that it takes a while to charge EVs—meaning a plugged-in customer will probably stay long enough to spend some money at the business.

If you work at a small business that may not have the funds for a charging station, you can suggest that your employer make electrical outlets available to EVs. Remember, the minimum infrastructure for an EV is a standard 110 V outlet. It won't charge quickly, but if you're at work all day, you can gain a good 40 to 50 km (25 to 30 miles) of range while it's plugged in.

Municipalities can fund EV infrastructure and install publicly accessible charging stations at convenient locations. This is something you can ask your council to do directly. However, an equally important role is to develop policies that allow private citizens to install EV stations on their own properties. This is particularly relevant for situations where people do not have a garage or driveway at their own home—and therefore would have no way to charge a car, aside from public chargers.

The city of Berkeley in California serves as a great model here. They adopted a curbside charging station policy, which allowed people to apply for a permit to install EV infrastructure on city property, if they did not have a garage or driveway. This has no material cost to the city—they just have to have the policy in place and allow the residents to do the rest. Higher levels of government can play a role in installing the more expensive fast-chargers, which allow EVs to be used to travel longer distances.

EVs are experiencing a chicken-and-egg problem, where people won't buy them until chargers are available, but people don't want to put in chargers until people are driving the EVs. It's time to solve this by making the infrastructure available. A little activation energy on the part of government will help spur widespread adoption of these vehicles, which will be better for all of us in the long run.

9. FUND RETRAINING PROGRAMS FOR FOSSIL FUEL WORKERS

No transition away from fossil fuels will be politically or practically feasible unless we answer a question that is like an elephant in the room: What are workers employed in the oil sector supposed to do in a society that needs far less of their product?

We need to be proactive about this. Climate justice must be social justice, and that means we can't leave behind thousands of people because they had the misfortune of being trained in the wrong industry. Here, government needs to step up and create a plan to retrain thousands of people in productive sectors that are not tied to fossil fuels. This will require a big investment in education.

Rather than being seen as a cost, this stands to be a tremendous opportunity. Workers in this sector are extremely qualified and have expertise that would make them excellent employees in a wide range of other sectors. Solar panels and wind farms are not going to install themselves, and an electrical grid of the future will not appear by magic. These will be put into place by the skilled hands of a giant workforce, and we should plan to help people get out of the fossil fuel sector and into one of these other industries.

With recent swings in global oil prices, retraining is more important than ever. In oil-producing provinces of Canada, thousands are currently out of work due to the recent drop in the price of a barrel of oil. Some people are calling for expansion of oil infrastructure, but this is precisely the wrong thing to do for our nation and for any nation interested in a sustainable future. Training a generation of world-class workers for the clean energy sector would be a natural transition, and it should be a priority of any government struggling to come up with sustainable pro-climate policies.

The solutions outlined in this chapter are simple in principle but can be difficult to enact. Nevertheless, if we allow ourselves to accept the true severity of the climate crisis, the rationale for taking these steps becomes clear. It's time to take action, and while we should certainly continue to study the best ways to move forward, we know enough now to begin to do what is needed to solve climate change. Now, it's a matter of political will.

Pursuing these solutions will make it easier for people to live by the carbon code. You have every right to demand all the things I've outlined here, even if you aren't an expert in any of these areas. Just as you don't have to be a civil rights lawyer to demand equal opportunity in society, you don't have to be an economist to demand that we adopt carbon taxes. You don't have to be an investment professional to understand the merits of divesting from fossil fuels. You don't have to be an engineer to understand that we're better off with charging stations rather than gas stations. And you don't have to be a climate scientist to understand that this is a problem we need to act deliberately, quickly, and effectively to solve.

There are those who will write op-eds saying that these solutions are not enough. Perhaps divestment isn't quite as good as carbon taxes—or perhaps EVs aren't as good as everyone switching to bicycles. That's not the point of this chapter—the point here is that you can argue passionately in favor of any or all of these items, and you'll be fundamentally correct. The devil may be in the details, but it's not your job to derive a complete solution to every aspect of the problem. Rather, living by the carbon code means taking action when you can and demanding actions of your leaders and superiors when the problem is bigger than you.

SUMMARY

→ *There are many policies that we should adopt that could help fight climate change.* Some are controversial among experts. Others are completely uncontroversial among experts but are not yet mainstream. Here are nine things we should already be doing. These are actions that would have a big effect on our ability to live by the carbon code and fight climate change.

1. Make polluters pay by adopting a carbon tax or cap-and-trade system.
2. Build tough regulations and close carbon outliers such as coal.
3. Explicitly prioritize climate change in decision making.
4. Halt all fossil fuel subsidies.
5. Subsidize clean energy.
6. Divest from the fossil fuel industry.
7. Develop cycling infrastructure.
8. Promote EV infrastructure.
9. Fund retraining programs for fossil fuel workers.

All nine of these actions are good policy, and there are examples of each being enacted somewhere in the developed world.

CHAPTER 10

BRINGING IT ALL TOGETHER

Climate change is truly insidious. It's as if some alien species designed a problem that humans are particularly ill suited to fix. Unlike wars, famines, or plagues, climate change doesn't strike the kind of unifying fear in us that sparks rapid society-wide action. It's slow, and it's not something that we can touch or feel. Even when we experience its consequences, such as through a major storm or an unseasonable wildfire, it's hard to tell to what extent climate change was really to blame for that specific catastrophe. It's like a specter that hangs over us, haunting our future.

Climate change is a trend, a mathematical reality that operates in terms of years and decades, not hours and minutes. It's a slow march toward an uncertain future, where the only sure thing is that life will be anywhere from a little to a lot harder than it is now. The poor will suffer most, as always, but the rich will be affected, too.

As a civilization, many of our collective innovations are designed to reduce the disorder that rules the existence of most species on the planet. We've tried to shape the planet to be more livable for us, and when we built our cities we chose habitats that best suited our way of life. Tremendous labor has gone into building our infrastructure, all of which has been underpinned by an assumption that the climate will be reasonably stable, as it has been for generations. As the climate warms, these rigid networks will be stressed. We will adapt as best as we can, but the reality is that if the seas rise as much as many scientists fear, then many of us will become climate refugees. Entire cities will become uninhabitable as they are swallowed by the sea, displacing millions. Populations will face disease and hunger. Many will die.

Some parts of the world will be inundated, while others will suffer droughts. Wars will be fought over scarce water and food. This is a human-made problem, and every single one of us who lives in an industrialized society bears part of the blame. Climate change is not caused by an obscure madman making the decision to oppress his people. It's caused collectively by all of us, through the millions of decisions we have to make just to go about our lives. Generally, the wealthiest of us are most to blame, but all of us contribute to the problem in our own way.

I am neither an optimist nor a pessimist. I prefer to confront reality head on, even if it's terrifying. I am tremendously hopeful, because I think we have a lot of power as individuals. Just as we're all part of the climate problem, we can all play a role in its solution. You really can be the hero in your own battle against climate change. But all heroes use rules, and that's where the carbon code can play a role. Even if you're a firm advocate for climate action, you're not exempt from managing your own carbon footprint. Deniers and climate villains will take every opportunity to undermine your work and efforts—let's lead by example, and remove the argument of hypocrisy they often deploy by taking care of our own actions first.

You may wonder why I didn't focus much on educating children about climate change. This was a deliberate choice on my part. It is incredibly important to educate them about the realities of the climate problem. However, while children are our future, adults are our present, and we don't have the luxury of waiting for the next generation to grow up and become leaders. By then, it will be too late. Furthermore, even when kids grow up and start wielding direct economic power, statistics demonstrate that they are unlikely to vote, at least initially. As difficult as it is, innovation and progress have to occur immediately, which means we have to engage with the adults of today and convince them of the merits of climate action. We need to set up carbon codes of conduct and engage people of all ages on the issue.

While it's important to educate yourself about a given topic, you don't need to have perfect knowledge of climate change to act on it. While this may sound odd for a scientist like myself to say, it is perfectly congruent with how we form opinions about most things in society. Take the computer that I wrote this book with as an example. I have no idea how this device works. I understand generally that there is a hard drive, a CPU, and a motherboard, but the physics behind how the computer actually processes data and the chemistry involved in producing all its components is beyond me. If I only acted on things I fully understood, I would not use a computer.

But we don't need to fully understand a topic to form an opinion. I'm not a forest manager, but I understand that we need to sustainably harvest trees to keep the forest intact while getting enough wood to build our homes. I'm not a doctor, but I understand that smoking is bad for me. I'm not a dietician, but I eat well to keep myself healthy.

I can do all these things because I trust that there are professionals out there who know as much about their fields of expertise as I do about marine conservation biology. That's why, when 99% of climate scientists tell me that climate change is a serious problem, I'm on board.

I propose the carbon code because it offers us a solution to the conundrum of climate change. It gives us a framework within which to make decisions about our carbon usage, while enabling us to participate in a carbon-intensive society. The code is scalable and can be implemented in your personal life, in any organizations to which you belong or lead, or even in government. It's a set of rules that remind us to always consider the needs of the atmosphere when we are making our decisions.

But how do you identify where to focus your efforts? In this book, we covered four priority areas that make up the majority of an individual's footprint. These were electricity, transportation, food, and long-distance travel.

For electricity, the best thing you can do is use less. This can be done by changing some of your behaviors—turning the furnace off before you leave the house and turning the lights off when you leave a room. Choosing efficient technology is the other branch of this. Replacing incandescent bulbs and other high-energy technologies with efficient units such as LEDs can take a big bite out of one's energy footprint.

But it also matters how your power is generated. Over this, you have some control. Solar panels are affordable and can be installed right onto your home. Energy storage, in the form of large batteries, can store this power for use when the sun isn't shining. Alternatively, net metering can allow you to sell power back to the grid so others can benefit from your generation. Any of these actions would be positive for the climate. And if you live in a jurisdiction that burns coal to make electricity, the relative benefit of taking these steps is far greater.

Transportation is another area where you have a lot of control. Switching from personal vehicles to public transportation is good, while using a bicycle is better. If you still have to use a personal vehicle, then carpooling can greatly reduce your per-capita emissions. And unless you drive a gas guzzler, keeping your car for longer before buying a new one keeps your footprint down as well, because it takes a lot of energy to build a new car. If you do buy a new car, then electric vehicles are the way to go. Whether your commute is short or long, there is a product available that can meet your needs, and while they cost more up front, the savings in fuel and maintenance make up for this over time.

When it comes to diet, hard choices must be made, but you don't have to be a vegan to have a positive impact on your climate footprint. Reducing the amount of beef and lamb you eat is the single best thing you can do. Beef must be reduced to a minor, negligible portion of the human diet because cows produce such huge amounts of methane, a greenhouse gas more potent than CO_2. Even though their methane is relatively short lived in the atmosphere compared to CO_2, it still represents an unacceptably high input of GHGs into our biosphere. Reducing other meats helps as well.

Travel is perhaps the most wicked of problems, especially for conservationists like myself. Airplanes produce incredible amounts of GHGs and burn lots of fossil fuels. The principle behind more-sustainable long-distance travel is to try to go low and go slow. Take trains, buses, or other means where possible, as an alternative to air travel. If you do have to fly, make the flight as direct as possible and accomplish at least two separate things on each trip. We should wring every drop of productivity out of the GHGs that we emit for flight and then invest in reputable offsets to rehabilitate the atmosphere.

Reduction, replacement, and refinement should guide our decisions in each of these four areas. Following these principles will enable us to lead by example. But once we do that, we actually have to be prepared to lead, and that's where winning the conversation comes in. Climate action is sensible action. It should be considered normal behavior, whereas unrestricted carbon pollution should be seen as abnormal.

The good news is that there are many examples of norms changing within our own lifetimes. Many governments have banned smoking in public buildings; this action would have been unthinkable only a few decades ago. Nowadays, we take it for granted that we should have some democratic say over who governs us. That is a relatively recent event and is still not the case in many countries. Even the consensus that women should be allowed to vote is only a few generations old.

So things do change, but they change through hard work and engagement. The impetus to change starts within a minority, and that minority slowly grows. This is where we're at with climate change—the minority is growing. But we don't have time to let this happen organically, and so people like you need to be ready to engage people on this issue. We need to make climate action normal.

Many of the actions we've covered in this book will have costs. Economically, some of them will be very expensive to deploy at the scale needed to solve the climate problem. Pragmatists will remind us that we still need oil and that there's only so much money to go around. But this book is not about pure pragmatism—it's about understanding that our economy and industries need to drastically restructure to accomplish the goal of carbon neutrality. Just because we need oil now doesn't mean we should allow that need to govern the way we build the society of the future. Any policy that reinforces our current dependence on oil is counterproductive, not just because it exacerbates climate change but also because it locks us further into an unsustainable system that threatens to harm us financially as the world shifts away from dirty energy. There is no such thing as ethical oil when our dependence on oil is inherently unethical.

Environmentally, we will pay costs as well. The clean technology I espouse in this book will require raw materials. An array of metals must be mined and processed in ways that are often very environmentally unfriendly to get them into a state ready for use in our low-carbon economy. We will need to make hard choices here. We must make them in an evidence-based manner, weighing the pros and cons of each decision within the greater goal of decarbonization.

I am not nuanced in my call to decarbonize, because the scientific consensus is that the severity of the climate problem far outweighs the costs of acting. We are surrounded by a media environment where people tell us that change is too difficult, too expensive, or too technologically challenging. Every year, each of these positions gets more wrong. Innovation doesn't just happen on its own—it happens when people push for it. So there is a role for aspiration, and there are few things more noble to aspire to than a carbon-neutral civilization.

But if no one is advocating for it then the change won't happen, and that's where you come in. The carbon code asks that you restructure the way you make decisions. We need to include our carbon footprint in our choices. Much like we consider quality, durability, and cost when we decide to buy a product, we must also consider the impact that product has on the biosphere.

You can be the agent of change, both in your own life and in the lives of those you work with. In chapter 8 we explored ways to discuss this issue. Simply asking your employer to adopt a carbon code of conduct will help the discussion begin. Posting a pro-climate viewpoint on a blog or on social media contributes in a similar way, and writing to your political representatives is better still. Collective action is built on millions of little actions, and you can be part of that solution.

In chapter 9, we looked at policies and actions that our political leaders should already be taking. Many of these are wildly controversial in the public discourse but are fundamentally sound policies. Pick one, or pick all of them—and make yourself a champion. Making change feels good! And what better way to give back to your community than to help bring it into the fold of sustainability?

When you're acting as an agent of change, approach it with joy. There is no right or wrong way to do it—some like to be militant, while others prefer to avoid confrontation. Some like to argue, while others enjoy bringing people together. It will take a full spectrum of types of people to move us toward a sustainable future and to bring us to the point where climate denial is socially unacceptable. The key is to do it in a way that gets you out of bed in the morning. If you're moving things in the correct direction, then far be it from anyone to criticize the tone or tenor of your actions.

You can be part of the solution, and I invite you to do so. Now that we've made the big decision to fight climate change, all the little decisions will fall into place. Let's close this book by making three more decisions before we go.

First, you will adopt the carbon code of conduct into your lifestyle. You will reduce, replace, and refine GHGs in your life. You will rehabilitate through carbon offsets when you do something that is particularly carbon intensive. You will live healthier, happier, and more secure as a result.

Second, you will do your best to convince others to adopt the carbon code. Every business, NGO, and government should have a publicly available carbon code of conduct. It should be standard practice to have one, and it should involve specific goals to manage the emissions in that workplace.

Third, you will contact your representatives in government, ask them what they are doing to solve climate change, and pressure them to do more. Invite them to adopt a carbon code of conduct in government.

In closing, let me state once again—anything you do to improve the climate is better than nothing. A whole lot of anything, even small things, amounts to a significant change. And through these actions, we will bring in a healthy, sustainable, equitable, and prosperous future. We can do it, and we will do it together.

NOTES

CHAPTER 2. SOLUTIONS START WITH YOU

1. *House of Commons Debates*, 41st Parl., 2nd Sess., No. 158 (December 9, 2014).

2. Tony Abbott. *Battlelines*. Melbourne: Melbourne University Publishing, 2010, p. 172.

3. Office of the Press Secretary. "President Bush Discusses Global Climate Change." Last modified June 11, 2001. http://georgewbush-white house.archives.gov/news/releases/2001/06/20010611-2.html.

CHAPTER 3. THE CARBON CODE OF CONDUCT

1. Ads of the World. "BC Hydro Power: Ridiculous Lights." Accessed January 31, 2016. http://adsoftheworld.com/media/tv/bc_hydro_power _ridiculous_lights.

2. Michael Lewis. "Obama's Way." *Vanity Fair*, September 30, 2012. http:// www.vanityfair.com/news/2012/10/michael-lewis-profile-barack-obama.

CHAPTER 4. ELECTRICITY

1. Google. "Project Sunroof." Accessed February 6, 2016. https://www .google.com/get/sunroof.

CHAPTER 8. WINNING THE CONVERSATION

1. Ezra Klein. "Al Gore explains why he's optimistic about stopping global warming." *Washington Post*, August 21, 2013. https://www.washingtonpost .com/news/wonk/wp/2013/08/21/al-gore-explains-why-hes-optimistic -about-stopping-global-warming/.

INDEX

Abbott, Tony, 34
Advanced Technology Vehicle
    Manufacturing Program, 192
air conditioning, 69, 74–75
aircraft, 143–144, 185
air pollution, 84–85, 155
Air Transport Action Group, 144
Ambio (journal), 135
Anderson, Kevin, 13
anti-conservationism, 38, 148,
    166–168
aquaculture, 137
Aquaculture Stewardship Council,
    137
aquifers, 15
Atwood, Margaret, 28
Australia, 20, 33–34

Bangladesh, 33
Banqiao Reservoir Dam, 90
Bell, Keith, 57
benign denialists, 166
Berkeley, 201
biofuels, 107–108; E85, 108
Bloomberg New Energy Finance
    Group, 98, 121
Brazil, 88, 108, 132; National
    Institute for Space Research, 88
British Columbia, University of, 195

Brown, Jerry, 42
Bullfrog Power, 82
Burch, Rex, 54
Bush, George W., 34

California, 16, 26, 108, 174, 184, 201
Campbell, Gordon, 108
Canada, 21, 33–34, 89, 172–174,
    183, 189; hydroelectric dams in,
    77, 89
cap and trade system, 181, 183, 203
Cape Verde, 15
carbon budget, 12, 36, 162–164, 181
carbon capture and storage (CCS),
    85, 186
carbon cycle, 10–11
carbon dioxide. *See* CO_2
carbon footprint, 33–34, 85–87; of
    electricity, 65–69, 74–75, 102; of
    food, 129–139; per capita, 33,
    154, 207; of transportation,
    101–107; of travel, 144, 151–155
carbonic acid, 18
carbon labels, 140
carbon tax, 138, 169, 173, 181–183
Carbon Trust, 140
cars. *See* personal vehicles
cement, 88
Chargepoint, 199–200

Chernobyl, 92–93
CH_4. *see* methane
Chile, 39
China, 35, 75, 84–85, 88, 90, 98, 126
citations, 23
civil disobedience, 176
Clean Air Act, US, 48
clean coal, 85, 186
Climatic Change (journal), 40
coal, 10, 36, 39, 65, 83–86, 102, 182
coal power plants, 39, 48, 75, 91, 150, 184–187
code of conduct, 54, 56; carbon, 47–61, 189, 210
Congressional Research Service, 191
conspicuous conservation, 39, 58, 61
Consultative Group on International Agricultural Research (CGIAR), 128–129
contrails, 144
Corporate Knights (magazine), 195
CO_2, 8–12, 31, 65, 96, 132, 143–144, 155, 163
CO_2e, 9, 33, 144, 151
CO_2-equivalent. *see* CO_2e
Credit Suisse, 33
Cruise Lines International Association, 155
cruises, 155–157
cycling, 121–123; infrastructure for, 197–199

dams: Banqiao Reservoir, 90; El Guri, 97; Muskrat Falls, 89; Three Gorges, 88
deforestation, 132, 181
Deloitte, 199
Democrats, 165
denialism, 22, 25, 44, 120, 161–168, 210
Department for Environment, Food, and Rural Affairs, UK, 151

depreciation, 107
diesel, 65, 84, 103–105, 126, 155–156
dilbit (diluted bitumen), 176
divestment, 194–196, 202
Dobzhansky, Theodosius, 28
drought, 15–16, 30, 41–44, 97, 163, 205

ecological efficiency, 130
ecology, 130
electric vehicles, 105–121, 124, 207; batteries for, 119; charging, 115–118; Chevrolet Bolt, 113; Chevrolet Volt, 112; Nissan Leaf, 113; Tesla Model S, 113–116, 119; Tesla Model 3, 113; Tesla Model X, 113
El Guri hydroelectric dam, 97
Energy Return on Investment (EROI), 93
Energy Star, 70, 102
Environmental Protection Agency (EPA), US, 70, 183
Environmental Research Letters, 16
Environmental Science and Technology, 133
Ethiopia, 33, 39
European Parliament, 191

feed conversion ratio (FCR), 131–133
fertilizer, 126–128, 134, 139–140; overfertilization, 127
fiduciary duty, 195
First Affirmative Financial Network, 195
fisheries, 135–136, 141, 191
flying, 144–148, 152, 157
Food and Agriculture Organization (FAO), 126, 135
food web, 136–137
Ford Motor Company, 43, 192

fossil fuels, 1, 38–43, 149–152,
    201–202; combustion of, 10–11;
    for electricity, 82–89; ethical
    use of, 47, 49, 52, 58, 61; and
    fertilizer, 139–140; mining of,
    120; for transportation, 103–113
fracking, 86–87, 190
Francis (pope), 42
Frid, Alejandro, 177
Friends of the Earth, 155–157
fuel cells, 107–109
fuel efficiency, 107, 170
Fukushima nuclear plant, 92–93, 95

gasoline, 47, 103–109, 111, 113–114
gas stations, 93, 104, 108, 111,
    116–118, 200
Gates, Bill, 42
George Washington University, 140
geothermal energy, 81–82
Getting There Green (Union of
    Concerned Scientists), 153
GHGs, 8–10, 31, 64–65, 84–88, 91,
    183–184
Global Warming Potential (GWP), 9
Gore, Al, 37, 161
GreenCarReports.com, 119
greenhouse gases. *see* GHGs
green parties, 39, 174
green technology, 37
Guardian, 106, 147, 196

Haber-Bosch reaction, 126, 139
Hansen, James, 22, 185
Harper, Stephen, 34
Harvard University, 195
heating, 69, 71, 74–76, 81, 101–102, 193
Heyns, Gerrit, 196
hydroelectricity, 77, 87–97, 102

Inconvenient Truth, An (Gore), 161
index funds, 196

India, 33
industrial revolution, 11, 36, 84, 119
information paralysis, 51, 61
Institute for Sustainable
    Resources, University College
    London, 188
insulation, 69, 193
internal combustion engine (ICE),
    107
International Atomic Energy
    Agency, 93
International Civil Aviation
    Organization, 185
International Monetary Fund
    (IMF), 31, 189
IPCC (Intergovernmental Panel on
    Climate Change), 11, 23–24, 64,
    128

Japan, 95, 109
*Journal of Personality and Social
    Psychology*, 168
justice, climate, 143, 150, 162, 186,
    201

Keystone XL pipeline, 187
Kharecha, Pushker, 185
kilowatt-hours (kWh), 66–68,
    72–73, 76–77, 80, 117–118
Ki-Moon, Ban, 42
Kinder Morgan, 176–177
Kiribati (islands), 15
Kyoto protocol, 34, 150

Lee, Paul, 114
lighting, 69–74; compact fluorescent
    (CFLs), 73–74; incandescent,
    72–74; light emitting diodes
    (LEDs), 72–74
Lindeman, Raymond, 130
locavores, 129
Los Angeles Times, 71

Maldives, 15
mangroves, 137
Marine Stewardship Council, 135, 142
Mars, 25
mass extinction: Permian, 18–19; sixth, 19
meat, 130–131; beef, 28–29, 32–33, 58, 60, 131–142, 207; chicken, 131–132; fish, 131; insect, 138; lamb, 132; pork, 132
medical research, 53–55
members of Parliament, Canada, 172, 181
mercury, 74, 85
metabolism, 127, 130
methane, 9, 21, 31, 88, 126, 132–134, 142, 207
methanogens, 132
Michigan State University, 127
microbes, 127, 134
Micronesia, 15
Minnesota, University of, 140
Morocco, 39, 77
Musk, Elon, 42
Muskrat Falls dam, 89
Myanmar, 137

National Academy of Sciences, US, 112
National Oceanic and Atmospheric Administration (NOAA), US, 99
natural gas, 74, 84–87, 126, 182
Nature, 19, 26, 188
Nature Climate Change, 97, 99
net metering, 79, 101, 207
New Democratic Party, Canada, 182
NIMBY (not in my backyard), 96
nitrous oxide. *see* N$_2$O
Nordhous, William, 12
N$_2$O, 9, 127, 134

nuclear energy, 90–97; fission, 91–97; fusion, 90–91; meltdown, 92–93
Nuclear Energy Institute, 91

Obama, Barack, 42, 58, 174, 185
oceans, 14–26, 110, 155–156; acidification of, 18; thermal expansion of, 15
offsets, 51, 148–150, 208; Gold Standard, 150; Plan Vivo, 150
oil, 27, 33–34, 39, 103–104, 110; spills of, 47, 110, 120, 187, 189
100-Mile Diet, The (Smith and MacKinnon), 128, 133
Overcoming Barriers to Deployment of Plug-In Electric Vehicles (National Research Council), 112
Oxfam, 33
Oxford University, 88

Palau, 15
Paris agreement, 41
payback period, 72–76, 81
Pentagon, 16, 31
permafrost, 21–22
personal vehicles, 104–107, 121, 124, 146, 207
Pew Research Center, 165
phantom load, 70–71
photosynthesis, 125
photovoltaic panels. *See* solar panels
pipelines, 83, 103, 107, 176
Plugshare.com, 118
positive advocacy, 168–170, 179
positive feedback, 21–22
power, 66–67
Powerwall battery, 79–80
precautionary principle, 163–164
Principles of Humane Experimental Technique (Russell and Birch), 54

Proceedings of the National Academy of Sciences, 14
Project Sunroof, 78

Qatar, 33
Quarmby, Lynne, 177

range anxiety, 115–118
regenerative braking, 111
religion, 38
Renewable Energy (journal), 78
Republicans, 165, 174
ruminants, 132–142
Russell, William, 54

Sagan, Carl, 25
salmon, Atlantic, farmed, 137
saltpeter, 126
Schwarzenegger, Arnold, 42, 108, 174
science, war on, 175
sea-level rise, 15, 22, 26
sequestration, 12
Simon Fraser University, 57, 176
Sinclair, Upton, 167
Site C dam project, 77
SkepticalScience.com, 22
social justice, 44, 65, 89, 150, 162, 184–186, 201
social license, 54–55
social media, 169, 210
solar panels, 16, 41, 67, 101–102, 192–193, 201; cost of, 89, 93; effectiveness of, 77–79
solar power, concentrated, 77
South Africa, 39, 196
Sovacool, Benjamin, 94
Stockholm Environment Institute, 150
subsidies: clean energy, 191–193; fossil fuel, 73, 95, 136, 189–190, 196, 203

sulphur dioxide (SO_2), 85
Sun Country Highways, 118
Superstorm Sandy, 15
swimming, 57
Swim to Win Playbook (Bell), 57
Syria, 16–17

teleconferencing, 145; Google Hangouts, 145; Skype, 145
10% rule, 130–131
teslacost.com, 114
Tesla Motors, 79, 115, 192; Powerwall, 79; Supercharger, 117
thermal tolerance, 19
thermoelectric power plants, 83–84, 97
thermostats, 75–76; programmable, 76; smart, 76, 100
350.org, 195
Three Gorges Dam, 88
tilapia, 137
time-of-day pricing, 80
Toronto, University of, 195
total cost of ownership (TCO), 106–107, 114
Toyota: Mirai, 109; Prius, 112, 114
tragedy of the commons, 36
transition fuel, 86–87
tree planting, 149
two-for-one principle, 147–148, 152, 157, 171
Tyndall, John, 9
Tyndall Centre for Climate Change Research, 87

Ukraine, 93
ultraviolet (UV) light, 8
Union of Concerned Scientists, 95, 153
United Arab Emirates, 33
United Kingdom, 13, 87, 141

United Nations Framework Convention on Climate Change, 41
United States, 33–34, 84, 103, 105, 108, 165, 191; politics of, 173–174

Vanity Fair, 58
veganism, 138, 207
vegetarianism, 137–139
Venezuela, 97

wildfires, 15, 163, 204
wind energy, 80–81, 100, 140, 149
Wind Energy Foundation, 80
World Health Organization (WHO), 26
World Nuclear Association, 94
World War II, 42

Yucca Mountain, 95